How to use this book

Welcome to *Space Science*. All the books in this set are organized to help you through the multitude of pictures and facts that make this subject so interesting. There is also a master glossary for the set on pages 58–64 and an index on pages 65–72.

The text is organized into chapters.

Capitals show key glossary terms. They are defined in the quick reference glossary.

Photographs and diagrams have been carefully selected and annotated for clarity. Captions provide more facts.

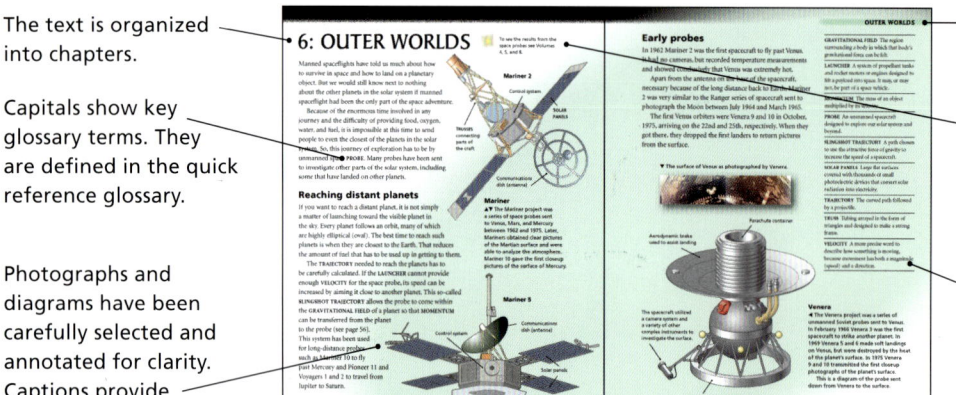

Chapter heading.

Links to related information in other titles in the *Space Science* set.

Quick reference glossary. All these glossary entries, sometimes with further explanation, appear in the master glossary for the set on pages 58–64.

 Atlantic Europe Publishing

First published in 2004 by
Atlantic Europe Publishing Company Ltd.

Copyright © 2004
Atlantic Europe Publishing Company Ltd.

All rights reserved. No part of this publication may be reproduced, stored in a retrieval system, or transmitted in any form or by any means—electronic, mechanical, photocopying, recording, or otherwise—without prior permission of the publisher.

Author
Brian Knapp, BSc, PhD

Art Director
Duncan McCrae, BSc

Senior Designer
Adele Humphries, BA, PGCE

Editors
Mary Sanders, BSc, and Gillian Gatehouse

Illustrations on behalf of Earthscape Editions
David Woodroffe and David Hardy

Design and production
EARTHSCAPE EDITIONS

Print
WKT Company Limited, Hong Kong

This product is manufactured from sustainable managed forests. For every tree cut down, at least one more is planted.

Space science – Volume 6: Journey into space
A CIP record for this book is available from the British Library

ISBN 1 86214 368 4

Picture credits
All photographs and diagrams NASA except the following:
(c=center t=top b=bottom l=left r=right)

Earthscape Editions 4–5, 6r, 7, 8b, 13tl, 13tc, 16l, 18, 19, 20, 27b, 31, 40t, 44–45, 46–47 *(all)*, 52t, 52b, 53bl, 54, 55, 56c, 56b, 57tr; *used with permission of Lucent Technologies Inc./Bell Labs* 29; *NASA Artist* 57tl, *and D. Seal* 57bl; *photo courtesy of Mrs. Robert Goddard/NASA* 15; *by permission of the Syndics of Cambridge University Library (from The Noonday Sun, Valerie Pakenham)* 6bl.

The front cover shows the Apollo 11 launch; the back cover, Gemini 6 and 7 meeting up in a rendezvous procedure.

NASA, the U.S. National Aeronautics and Space Administration, was founded in 1958 for aeronautical and space exploration. It operates several installations around the country and has its headquarters in Washington, D.C.

2

CONTENTS

1:	**INTRODUCTION**	4–13
	The first steps	6
	From arrows to rockets	6
	Controlling the projectile	8
	The nature of launch vehicles	9
	Launcher design	10
	Propulsion systems	10
	Engines and motors	11
	How launcher engines and motors work	12
2:	**ROCKETRY**	14–21
	The rocket pioneers	14
	Converting military launchers	18
	Modern launchers	20
3:	**THE SPACE RACE**	22–29
	Sputnik	22
	The American challenge	24
	Explorer and Vanguard	24
	Discoverer	26
	Tiros	27
	Telstar	28
4:	**MANNED SPACEFLIGHT**	30–37
	Vostok	33
	Mercury	34
5:	**TO THE MOON**	38–51
	Gemini	38
	The Apollo project	43
	To the Moon—carefully	47
6:	**OUTER WORLDS**	52–57
	Reaching distant planets	52
	Early probes	53
	From Pioneer to Cassini	54
	SET GLOSSARY	58–64
	SET INDEX	65–72

▲ The augmented target docking adapter (ATDA) as seen from the Gemini 9 spacecraft. The ATDA and Gemini 9 are 60 meters apart. Failure of the docking adapter protective cover on the ATDA to fully separate prevented the docking of the two spacecraft. The ATDA was described by the Gemini 9 crew as an "angry alligator."

1: INTRODUCTION

SPACE, a distant mystery.... And so it has seemed to people through the centuries. Until the last half century anyone could only look, for there was never a chance of reaching out into space. Even Earth-bound telescopes are very limited in what they can help us see, partly because the Earth's ATMOSPHERE obscures distant vision, and partly because many objects in space (such as the MOON) show us only one side all year.

The development of SPACECRAFT has changed all our perspectives on space. No telescope picture comes remotely close to the sensation we all feel when we see an astronaut put a foot down on the Moon, or when a PROBE takes pictures close to a moon on the edge of our SOLAR SYSTEM, revealing beauty that could not have been imagined.

▲▶ Although we are very familiar with seeing rocket launches at a distance (*above*), it is only when we come face to face with a launcher, such as Saturn V at the Kennedy Space Center (*right*), that we can start to appreciate the sheer vastness of the power needed to get into space.

4

INTRODUCTION

ATMOSPHERE The envelope of gases that surrounds the Earth and other bodies in the universe.

MOON The natural satellite that orbits the Earth.

PROBE An unmanned spacecraft designed to explore our solar system and beyond.

SOLAR SYSTEM The Sun and the bodies orbiting around it.

SPACE Everything beyond the Earth's atmosphere.

SPACECRAFT Anything capable of moving beyond the Earth's atmosphere. Spacecraft can be manned or unmanned. Unmanned spacecraft are often referred to as space probes if they are exploring new areas.

The first steps

Two challenges face anyone trying to get a spacecraft into space. The first is how to escape from the Earth's GRAVITY, and the second is how to survive in space. For manned spaceflight there is a third, and perhaps the biggest challenge of all: how to get back to the Earth safely. This book deals with all these aspects up to the time of APOLLO and VOYAGER in the early 1970s.

From arrows to rockets

The first steps into space do not begin with building a spacecraft. They begin on a much simpler level with a little science. Then, once we grasp some basic science ideas, it becomes possible to move toward the technological challenges of space travel.

Have you ever looked closely at an arrow? It is a long shaft, sharpened at one end and with a set of feathers, called flights, at the other. The arrow is long, thin, and pointed so that it can fly through the air with the minimum of AIR RESISTANCE. But a shaft on its own is uncontrollable. The flights at the end are what make it fly in a straight line.

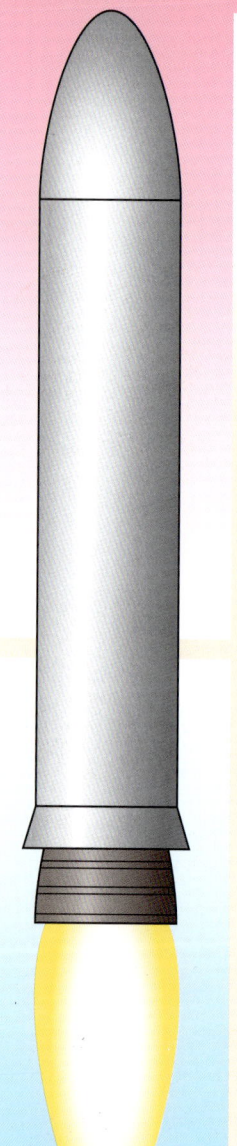

▼ A rocket is a reaction engine.

◄ Traditional use of a bow and arrow.

For all developments that have occurred with the Space Shuttle and International Space Station see Volume 7: *Shuttle to Space Station*. For modern satellite missions see Volume 8: *What satellites see*.

INTRODUCTION

Now think about a rocket firework. A rocket of this kind is like a powered arrow. It is designed like a shaft so that it will meet as little air resistance as possible. But it is not shot by a bow; it has its own **PROPELLANT** (fuel) on board. Although this means that it no longer depends on an external propulsion system and can go much farther, the disadvantage is that it cannot be as slim or light as an arrow.

To keep down the air resistance created by its larger size, the rocket has a nose cone. That pushes the air aside and around the rocket's sides, and is an example of **AERODYNAMIC** styling.

The propellant inside the body of a rocket firework is a solid mixture of chemicals (called solid fuel) that is a version of gunpowder. This chemical mixture (72% nitrate, 24% carbon, and 4% sulfur) is adapted so that it will burn fiercely but not so fast that it explodes.

When the blue fuse is lit, the solid gunpowder changes into a gas, expanding and causing an increase in gas **PRESSURE**. The strong casing keeps these gases from escaping anywhere but from the bottom.

As the gases push against the air, an equal and opposite **REACTION** is produced that lifts the firework off the ground. This is called a reaction motor (see also pages 12–13).

The explosive nature of a firework propellant makes it **ACCELERATE** very quickly and keep accelerating until it has used up all the propellant. But there is no way of closely controlling the rate at which the fuel is burned or guiding the rocket accurately. When the rocket has used up all of its propellant, it falls back to the ground and usually breaks up.

▶ The components of a firework rocket.

ACCELERATE To gain speed.

AERODYNAMIC A shape offering as little resistance to the air as possible.

AIR RESISTANCE The frictional drag that an object creates as it moves rapidly through the air.

APOLLO The program developed in the United States by NASA to get people to the Moon's surface and back safely.

GRAVITY The force of attraction between bodies.

PAYLOAD The spacecraft that is carried into space by a launcher.

PRESSURE The force per unit area.

PROPELLANT A gas, liquid, or solid that can be expelled rapidly from the end of an object in order to give it motion.

REACTION An opposition to a force.

THRUST A very strong and continued pressure.

VOYAGER A pair of U.S. space probes designed to provide detailed information about the outer regions of the solar system.

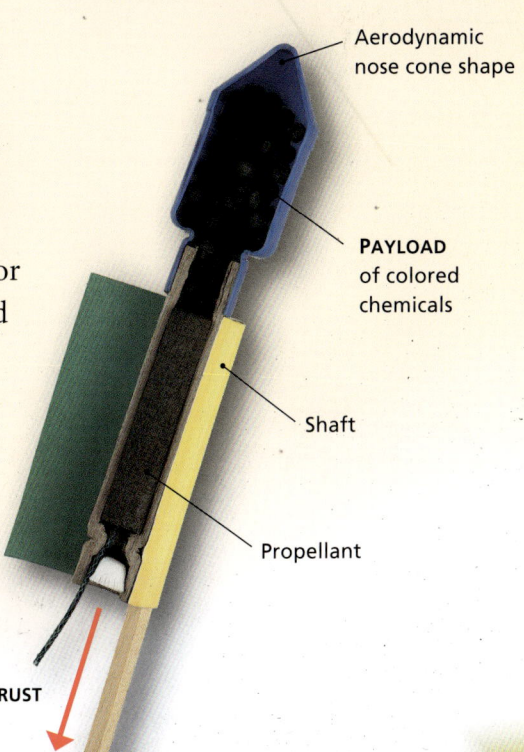

Aerodynamic nose cone shape

PAYLOAD of colored chemicals

Shaft

Propellant

THRUST

7

Controlling the projectile

In the arrow and the firework we have the science needed to get people to the Moon and beyond. We know that we need some **PROPELLANT** to make the **PROJECTILE**—that is what any **ROCKET** is—move very quickly. It is always counteracted by the Earthbound force of **GRAVITY**.

So to leave the Earth and get into space, the propellant has to produce more **THRUST** than the force of gravity for long enough to get the rocket and its cargo, such as a **SATELLITE** or a manned spacecraft—called the **PAYLOAD**—to beyond the **GRAVITATIONAL PULL** of the Earth.

Most of this time the rocket is moving through an atmosphere of gases that are constantly applying **FRICTION** (drag) to its surface. Anything moving fast through the air will heat up because of this surface friction. The buildup of heat in a spacecraft may reach dangerous levels, especially on reentry into the Earth's atmosphere.

So, first and foremost the rocket needs to be streamlined and well protected against frictional heating. But it also has to be strong to hold the weight of the fuel inside. Also, as the spacecraft goes up, the air pressure on the casing gets less, causing the casing to tend to burst apart.

The rocket also needs some way of controlling the rate at which the propellant is used up. Then it needs some means of steering. And finally, it needs a means of getting back safely to the Earth's surface, so that it doesn't just smash up as it returns.

▼ If a projectile is thrown with a low velocity, it will soon fall to the ground. As the velocity is increased, it will go farther before it falls. At a great enough velocity it will go into orbit (4).

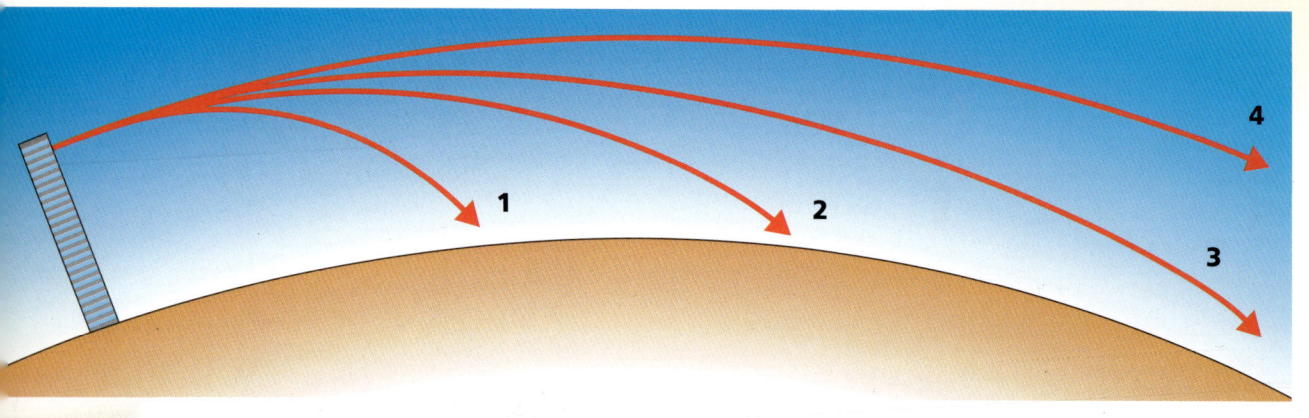

INTRODUCTION

◀ Stage 1 of a launcher falls away, its fuel having been used up.

The science behind what has to be done is very straightforward in principle. But changing the science into technology that works—now that's a very different matter, and one that has challenged people for over a century and is still challenging them today.

The nature of launch vehicles

A rocket is really a simple all-in-one projectile. Modern spacecraft are much more sophisticated, consisting of a number of separate items connected together. A **LAUNCH VEHICLE** (or launcher) is what most of us think of as a rocket. It is the **PROPULSION SYSTEM** that allows its cargo to **ORBIT** the Earth or to leave the Earth altogether. It is by far the biggest part of the whole assembly and may weigh well over 90% of the total.

The payload of the launcher is the spacecraft. The spacecraft may be a satellite, or it may be a manned spacecraft. If an unmanned spacecraft is going to travel long distances across the solar system or beyond to newly explored areas, it may be called a space **PROBE**.

The job of the launcher is to get the payload either into orbit around the Earth or to send it beyond the gravity of the Earth.

Getting a payload into space requires twice as much fuel as putting it into orbit, so the payload for a space journey can only be half as heavy as the payload on the same launcher going into orbit.

It takes much more fuel to go directly into space than to go first into orbit and then leave orbit for space. So all spacecraft follow a curving path, or **TRAJECTORY**, that first takes them into orbit. The least a launcher must do, therefore, is be able to lift its payload into an orbit around the Earth. To do this, the launcher must achieve a **VELOCITY** of 7.5 kilometers a second. That was not possible until 1957.

FRICTION The force that resists two bodies that are in contact.

GRAVITY/GRAVITATIONAL PULL The force of attraction between bodies. The larger an object, the more its gravitational pull on other objects.

LAUNCH VEHICLE A system of propellant tanks and rocket motors or engines designed to lift a payload into space. It may, or may not, be part of a space vehicle.

ORBIT The path followed by one object as it tracks around another.

PAYLOAD The spacecraft that is carried into space by a launcher.

PROBE An unmanned spacecraft designed to explore our solar system and beyond.

PROJECTILE An object propelled through the air or space by an external force or an on-board engine.

PROPELLANT A gas, liquid, or solid that can be expelled rapidly from the end of an object in order to give it motion.

PROPULSION SYSTEM The motors or rockets and their tanks designed to give a launcher or space vehicle the thrust it needs.

ROCKET Any kind of device that uses the principle of jet propulsion, that is, the rapid release of gases designed to propel an object rapidly.

SATELLITE A man-made object that orbits the Earth.

THRUST A very strong and continued pressure.

TRAJECTORY The curved path followed by a projectile.

VELOCITY A more precise word to describe how something is moving, because movement has both a magnitude (speed) and a direction.

Going into space
▲ The payload is a tiny front-end compartment of the space vehicle (Apollo 11).

Launcher design

Launchers are always built in detachable stages because it is important to lose as much unnecessary weight as quickly as possible during the journey. If a single stage was used, the tanks would be partly empty for much of the journey, adding the weight of the casing for no gain.

The launcher and its payload weigh most when they have a full fuel tank, and most fuel is used up simply getting the assembly off the ground and through the lower atmosphere. That is why the tanks are the biggest part of any spacecraft, and why some of them quickly become empty and are dropped off within a couple of minutes of takeoff.

Propulsion systems

A PROPULSION SYSTEM can be considered the power unit or engine of the rocket. It burns the propellant or fuel. But launcher propulsion systems are not like the engines we use in cars.

Rocket propulsion systems are special forms of the jet engine. The main difference between spacecraft and aircraft jet engines is that normal jet engines use the air of the atmosphere to burn their fuel, so they do not have to carry around a supply of oxygen in a large tank. However, using air does have a big disadvantage: Burning needs oxygen, and the amount of oxygen that can be used in burning fuel depends on how much oxygen is in the air. That limits the thrust the engine can provide. More thrust can be achieved by using a more concentrated supply of energy, even on the ground. However, launchers have to operate in space, where there is no oxygen at all, and so they have to carry their oxygen supply with them in the form of huge tanks of liquefied oxygen gas.

Propulsion systems can look crude, but they are not. They do have to be as simple as possible, so that they are reliable. They also have to be able to withstand enormous forces and the flow of searingly hot, CORROSIVE SUBSTANCES.

INTRODUCTION

Engines and motors

Launchers use either solid propellants or liquid propellants (fuels) in their propulsion systems. The term **ROCKET MOTOR** is used if the propulsion system burns solid fuel and **ROCKET ENGINE** if it burns liquid fuel.

Solid fuels produce more thrust, but the rate of fuel usage cannot be closely controlled; liquid fuels produce less thrust, but are controllable in that the amount of fuel can be changed, just as we might change how far we press down on the gas pedal of a car.

Since both types of fuel have their own advantages and disadvantages, both are usually used. Solid fuels often power early launcher stages, for example, getting the launcher off the ground. Liquid fuels propel the last phases of ascent and aid maneuvering.

The rate at which a fuel of any kind will burn depends on how quickly oxygen reaches it (as you notice when you increase the draft to a fire). In this role oxygen is called an **OXIDIZER**. The amount of oxygen in air is small, and the concentration of oxygen decreases as you go up through the atmosphere. In addition, if you simply suck air in from the surroundings, you have little control over how the fuel burns. So, the best solution is not to rely on air at all and use oxygen (as, for example, in a liquefied gas).

Kerosene is a common liquid fuel, stored in one tank strapped to the launcher, with liquid oxygen in a second tank. Early launchers used this system. Other chemicals include hydrazine as fuel and nitrogen tetroxide, instead of oxygen, as the oxidizer. These chemicals ignite spontaneously when they mix, avoiding the need for an igniter as with kerosene and oxygen. Engines that ignite spontaneously are called hypergolic engines.

There are alternatives to liquid and solid fuels. One important one is hydrogen. Hydrogen is a gas. In its gaseous form it would take up a huge volume (think of the airships of the past), so it has to be compressed to liquefy before use.

CORROSIVE SUBSTANCE Something that chemically eats away something else.

OXIDIZER The substance in a reaction that removes electrons from and thereby oxidizes (burns) another substance.

PROPULSION SYSTEM The motors or rockets and their tanks designed to give a launcher or space vehicle the thrust it needs.

ROCKET ENGINE A propulsion system that burns liquid fuel such as liquid hydrogen.

ROCKET MOTOR A propulsion system that burns solid fuel such as hydrazine.

The weight equation

To launch a space vehicle requires an enormous fuel-to-payload weight ratio. In the Apollo mission a huge Saturn V rocket was used to launch the small command module and Moon lander. The equation remains true. You can see it here, for example, with the Space Shuttle (figures below are approximate).

Orbiter empty:	75,000 kg
External tank empty:	35,000 kg
Two solid rocket boosters empty:	170,000 kg
Fuel	
External tank:	700,000 kg
Solid rocket boosters:	1,000,000 kg
Total fuel:	1,700,000 kg

Total weight at launch, including tanks, orbiter, and its payload: about 2,000,000 kg

Fuel alone weighs over 20 times the weight of the payload. Fuel plus tanks weigh nearly 30 times the payload.

How launcher engines and motors work

◀ Detail of an engine.

COMBUSTION CHAMBER A vessel inside an engine or motor where the fuel components mix and are set on fire, that is, they are burned (combusted).

GIMBALS A framework that allows anything inside it to move in a variety of directions.

GYROSCOPE A device in which a rapidly spinning wheel is held in a frame in such a way that it can rotate in any direction. The momentum of the wheel means that the gyroscope retains its position even when the frame is tilted.

LAWS OF MOTION Formulated by Sir Isaac Newton, they describe the forces that act on a moving object.

REACTION An opposition to a force.

Space engines are **REACTION** engines. They work by sending gases in one direction and benefiting from the reaction that happens as a result.

To understand this, think about inflating a balloon and then letting go of it. The gas in the balloon is under pressure. It is released through a small opening that makes the gases move as fast as possible. The movement of the balloon occurs because, as Newton's **LAWS OF MOTION** state, "For every action there is an equal and opposite reaction." The gases expand and move away from the balloon. This sets up an equal and opposite (reaction) thrust in the balloon, making it move in the opposite direction from the gases—whether in the atmosphere or in space.

By burning the propellants and forming gases that are expelled from the engines or motors at high speed, launchers develop enormous thrust. By controlling the exhausts using **GYROSCOPES** and cradles called **GIMBALS**, it is possible to steer them.

The mixing of fuel and oxidizer happens at such a rate that inside the **COMBUSTION CHAMBER** there is something similar to an explosion. Keeping control of the rate of fuel burning is a difficult problem. You need to control the rate at which fuels enter the combustion chamber and turn to gases, not just let the liquids or solids run together under gravity. So, you need to pump the gases and design pumps that will deliver huge amounts of gas very quickly. To make this even more efficient, liquid fuels are first mixed and pressurized before they reach the main combustion chamber. There they burn to create a high-pressure and high-speed stream of hot gases that flow through a nozzle at speeds of 160,000 km/hr.

INTRODUCTION

Fuel tanks

Pumps draw fuel into combustion chamber

Valves regulate the fuel flow

Combustion chamber

Nozzle directs the thrust

▲ This diagram shows how fuel and oxidizer from two separate tanks are fed, via pumps, to the combustion chamber. Modern systems are much more sophisticated but still work on the same principle.

◀ Solid fuel motors are filled with chemicals, and a central shaft is left clear. The fuel around this shaft is then ignited. The motor burns from the inside outward. There is no means of controlling the rate of burning.

▲▼ An engine being fired up during a test.

▲ The three engines of the Space Shuttle showing housing in steering gimbals.

13

2: ROCKETRY

When we look at rockets today (now called LAUNCHERS), we see a highly sophisticated machine. But these launchers have been developed over a long time.

Interest in ROCKETRY goes back many centuries, but its practical development is only about a century old. It is curious now to look at photographs of some of the early rockets, which were simple and crude. The ones developed by Robert Goddard in America in the 1920s are typical (other centers of rocketry included the Soviet Union and Germany). What developers were trying to do was establish the scientific and technological principles, or ground rules, from which everything else would follow.

LAUNCHER A system of propellant tanks and rocket motors or engines designed to lift a payload into space. It may, or may not, be part of a space vehicle.

ROCKETRY Experimentation with rockets.

The rocket pioneers

It took huge efforts by some rocket pioneers to get rocketry under way. Three people dominated those early years— Robert H. Goddard of America (he gave his name to the Goddard Space Flight Center), Konstantin Eduardovich Tsiolkovsky of Russia, and Hermann Oberth of Germany.

Robert H. Goddard

Goddard was a research scientist (and an inventor) with a flair for technology. In 1907 he began his experiments using solid fuel. He chose to test his rocket in the basement of the physics building at Worcester Polytechnic Institute!

It was Goddard who, in 1914, also developed the idea of using liquid fuel. He built and successfully tested the first rocket using liquid fuel on March 16, 1926, at Auburn, Massachusetts. This feat was the rocketry equivalent of the first aircraft flight by the Wright brothers.

The difference between aircraft development and rocketry was that governments of the time could see many possible uses for aircraft, but none for rockets. So, at first Goddard's ideas were ridiculed.

▼▶ Robert Goddard with his double acting engine rocket in 1925. This rocket has an engine that uses a separate pump for each propellant. The idea of combining both pumps into a single unit made pumps more reliable and hence marked a significant advance in controlled rocket design.

14

ROCKETRY

15

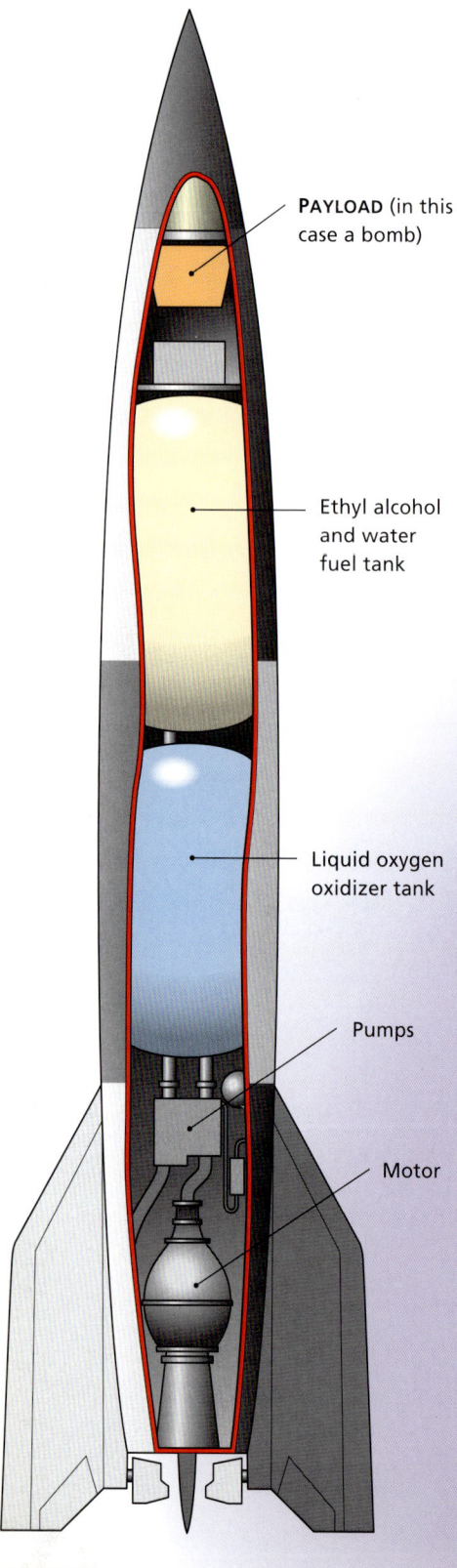

▼ The V2 rocket.

In Germany, however, his ideas were taken seriously, and the German Rocket Society was formed in 1927. Very soon afterward the German military discovered the potential of rockets, and in 1931 it set up a rocket program. From this the Germans developed both the V1 (Vengeance Weapon 1) and the V2 rockets as flying bombs (**BALLISTIC MISSILES**), which were used with devastating results in World War II.

The Germans were able to use Goddard's principles and develop them to great effect. They created gyroscopic control, steering by means of vanes (fins), a means of attaching the motor in **GIMBALS** so that it could be used for steering, and power-driven fuel pumps.

Konstantin Tsiolkovsky

Konstantin Tsiolkovsky (1875–1935) was a Russian pioneer, using wind tunnels to achieve good **AERODYNAMIC** design for rockets.

Hermann Oberth

Hermann Oberth (1894–1989) was a German and one of the first to consider the nature of weightlessness. He also designed an early long-range, liquid-propellant rocket. The first rocket designed by him was launched on May 7, 1931, near Berlin. During World War II he worked for his former assistant, Wernher von Braun, on V1 and on V2 rockets.

Wernher von Braun

Wernher von Braun (1912–1977) was stimulated into rocket science by his contact with Hermann Oberth. He helped Oberth with liquid-fueled **ROCKET ENGINE** tests, later working for the German military.

▶ Trial launch of the V2.

ROCKETRY

▼ Wernher von Braun with President John Kennedy and a model of a Saturn launcher.

By the end of World War II German scientists like von Braun were world leaders in the rocketry field, still making good use of the pioneering work of Robert Goddard.

After the war von Braun and his group were transferred to the U.S. Army. By 1952 von Braun was technical director (then chief) of the U.S. Army ballistic-weapons program and later became a U.S. citizen. He was important in developing the spaceflight program.

It was von Braun's team that launched the first U.S. satellite, Explorer 1, on January 31, 1958. With the founding of **NASA** von Braun was transferred to the George C. Marshall Space Flight Center, where he became director and supervised the development of large space launch vehicles, such as Saturn I, IB, and V (see page 20).

AERODYNAMIC A shape offering as little resistance to the air as possible.

BALLISTIC MISSILE A rocket that is guided up in a high arching path; then the fuel supply is cut, and it is allowed to fall to the ground.

GIMBALS A framework that allows anything inside it to move in a variety of directions.

NASA The National Aeronautics and Space Administration.

PAYLOAD The spacecraft that is carried into space by a launcher.

ROCKET ENGINE A propulsion system that burns liquid fuel such as liquid hydrogen.

Converting military launchers

Both the Soviet Union and the United States already had considerable rocketry experience because they had both been building intercontinental ballistic missiles (ICBMs) so that they could launch **ATOMIC WEAPONS** at one another in the event of an attack. Both countries used modified ICBMs to launch their early spacecraft.

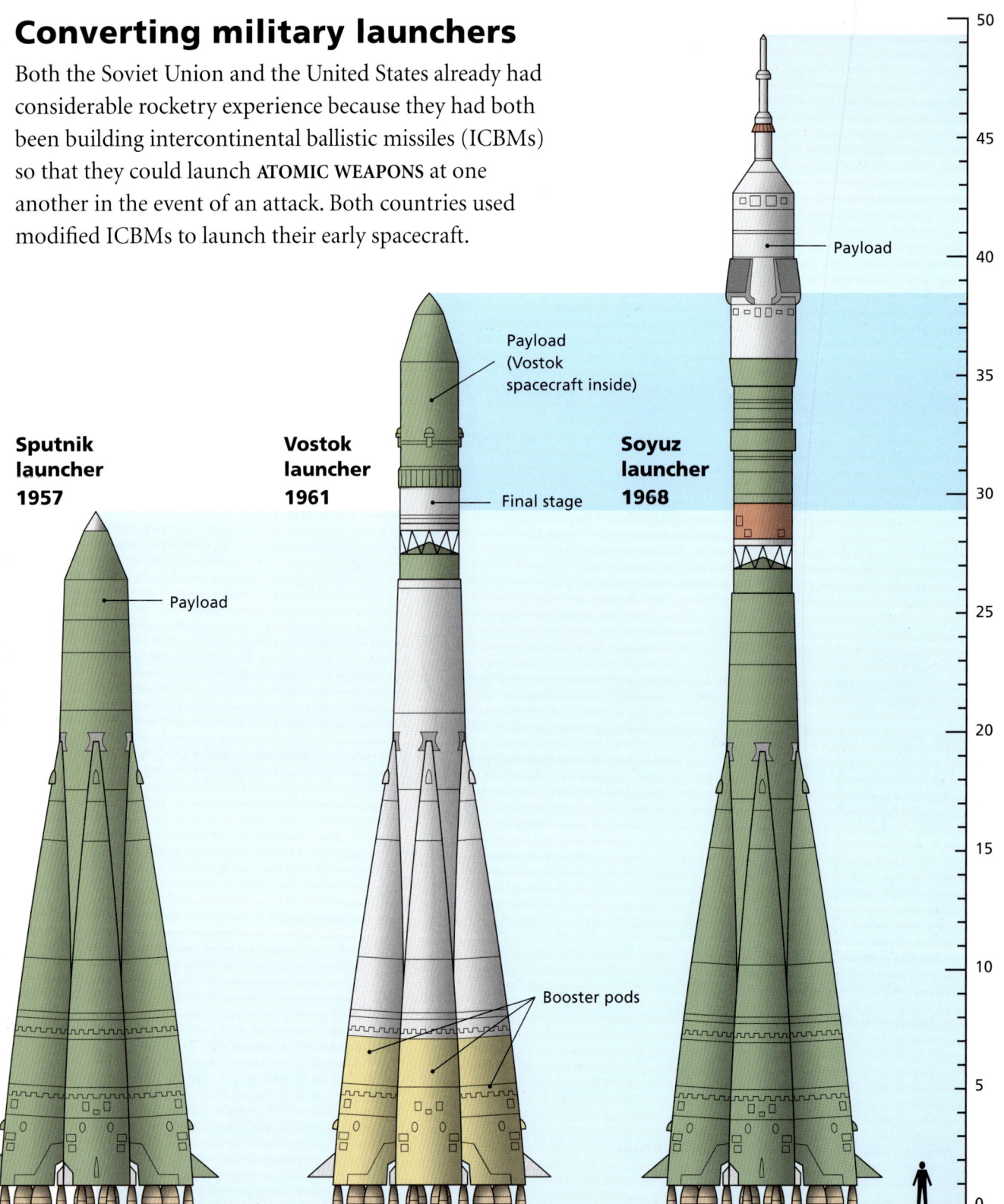

▲ Early Russian launchers were generally more powerful than their American counterparts. They were made bigger in order to launch heavier payloads into orbit.

ROCKETRY

▲ The wide variety of launchers now on display at the Kennedy Space Center.

Other countries did not have such launchers and so were not able to join in the SPACE RACE.

The first launch vehicles used by the Soviet Union had two stages with four drop-away BOOSTER PODS. Each booster pod contained four rocket engines. The main launcher core also had four engines. That gave the launcher a total of 20 engines. The pods also contained the tanks for the propellant (fuel), which used liquid oxygen and kerosene.

The first American launchers, Jupiter and Vanguard, also used liquid fuels in the lower part of the launcher and solid propellant in the upper stages.

In the 1960s another ICBM was modified for use, called Titan. Another reused ICBM was called Thor, and yet another was called Atlas.

ATOMIC WEAPONS Weapons that rely on the violent explosive force achieved when radioactive materials are made to go into an uncontrollable chain reaction.

BOOSTER POD A form of housing that stands outside the main body of the launcher.

SPACE RACE The period from the 1950s to the 1970s when the United States and the Soviet Union competed to be first in achievements in space.

Modern launchers

It was not until the **APOLLO** Moon landing program that launchers were designed specifically for nonmilitary space use. The most famous of these launchers was called Saturn. Saturn 1B was made of two stages, and Saturn V had three stages. It was Saturn V that was eventually used for the lunar (Moon) missions.

Saturn V was 110.6 meters high and weighed over 2,700 tonnes when it was launched. Its payload was impressive for the time. It could put 104 tonnes of payload into Earth **ORBIT** and could propel 45 tonnes at a **VELOCITY** great enough to escape from the Earth.

By the 1960s other countries had developed launchers capable of putting **SATELLITES** (but not manned spacecraft) into orbit. France, China, and the United Kingdom all developed such launchers. European launch technology was eventually pooled under the heading of the European Space Agency (ESA), and the launcher Ariane was developed. Ariane is a three-stage vehicle. It uses solid **PROPELLANTS** in its first two stages but uses a cryogenic engine (an engine that burns hydrogen and oxygen fuels) in the third stage. It can launch payloads of up to 4.7 tonnes, which means it can launch two "standard" satellites at a time.

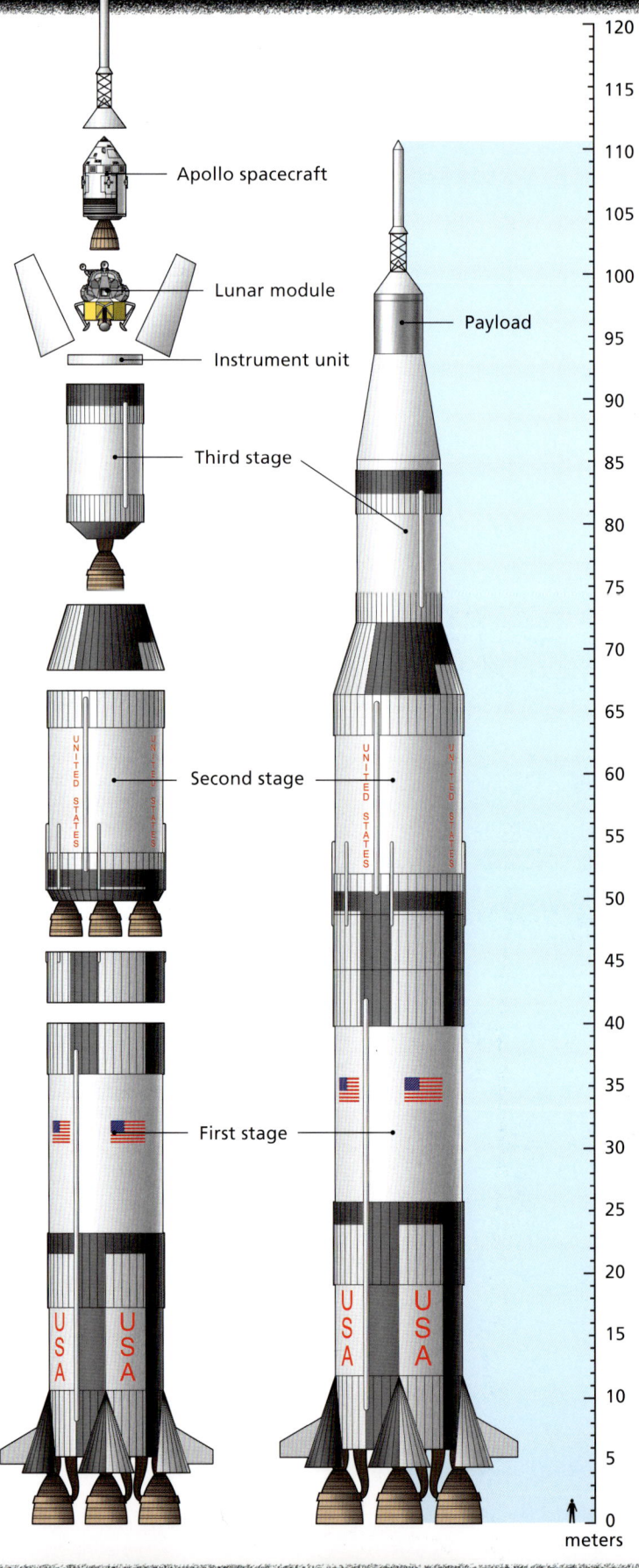

▶ The giant Saturn V launcher.

20

ROCKETRY

▲ The Cassini orbiter being carried by the launcher Titan IVB/Centaur. Cassini and its Huygens probe were launched in 1997 to begin their long journey to Saturn (see page 57).

APOLLO The program developed in the United States by NASA to get people to the Moon's surface and back safely.

PROPELLANT A gas, liquid, or solid that can be expelled rapidly from the end of an object in order to give it motion.

SATELLITE A man-made object that orbits the Earth.

VELOCITY A more precise word to describe how something is moving, because movement has both a magnitude (speed) and a direction.

21

3: THE SPACE RACE

The development of vehicles capable of traveling into space was a complex affair that took decades. As the last chapter showed, much was learned during World War II, but the idea of leaving something useful—a **PAYLOAD**—in space still had to be realized. There was no need to design a payload that was **AERODYNAMIC**, because in space there is no air and so no **AIR RESISTANCE**. That is why the first satellites looked very different from any rocket or aircraft.

The development of these machines was not simply a matter of science and technology. During this period there was also intense political competition between the two superpowers, the United States and the Soviet Union, to be the first to master this new and powerful technology. It became known as the **SPACE RACE**.

The political arena tended to change, or distort, what scientists and technologists might otherwise have done. We will see that in this chapter.

Sputnik

The first object put into space was Sputnik 1. It was launched on October 4, 1957, by the then Soviet Union. It looked rather like a metal ball with skewers sticking out of it.

The launcher carried the **SATELLITE** to a speed of 7.99 km per second, thus allowing it to escape the Earth's **GRAVITY**. As it entered **ORBIT**, a nose cone came away, revealing a further nose cone, which was also released. Sputnik, being spring-loaded, was then ejected from its launcher. Four aerials sprang out, each 3 m long, and they provided the communications for the two transmitters.

It was possible to receive the signals all over the world, including in schools. The Morse-codelike (bleep, bleep) signal was heard around the world. It marked the start of the Space Age in the most dramatic fashion. Sputnik weighed 82 kg and at its highest orbital point reached 939 km above the Earth's surface. Its orbit took 1 hour 36 minutes.

AERODYNAMIC A shape offering as little resistance to the air as possible.

AIR RESISTANCE The frictional drag that an object creates as it moves rapidly through the air.

DOPPLER EFFECT The apparent change in pitch of a fast-moving object as it approaches or leaves an observer.

FREQUENCY The number of complete cycles of (for example, radio) waves received per second.

GLOBAL POSITIONING SYSTEM A network of geostationary satellites that can be used to locate the position of any object on the Earth's surface.

GRAVITY The force of attraction between bodies.

IONOSPHERE A part of the Earth's atmosphere in which the number of ions (electrically charged particles) is enough to affect how radio waves move.

OPTICAL Relating to the use of light.

ORBIT The path followed by one object as it tracks around another.

PAYLOAD The spacecraft that is carried into space by a launcher.

RADIO WAVES A form of electromagnetic radiation, like light and heat. Radio waves have a longer wavelength than light waves.

SATELLITE A man-made object that orbits the Earth.

SENSOR A device used to detect something.

SPACE RACE The period from the 1950s to the 1970s when the United States and the Soviet Union competed to be first in achievements in space.

THE SPACE RACE

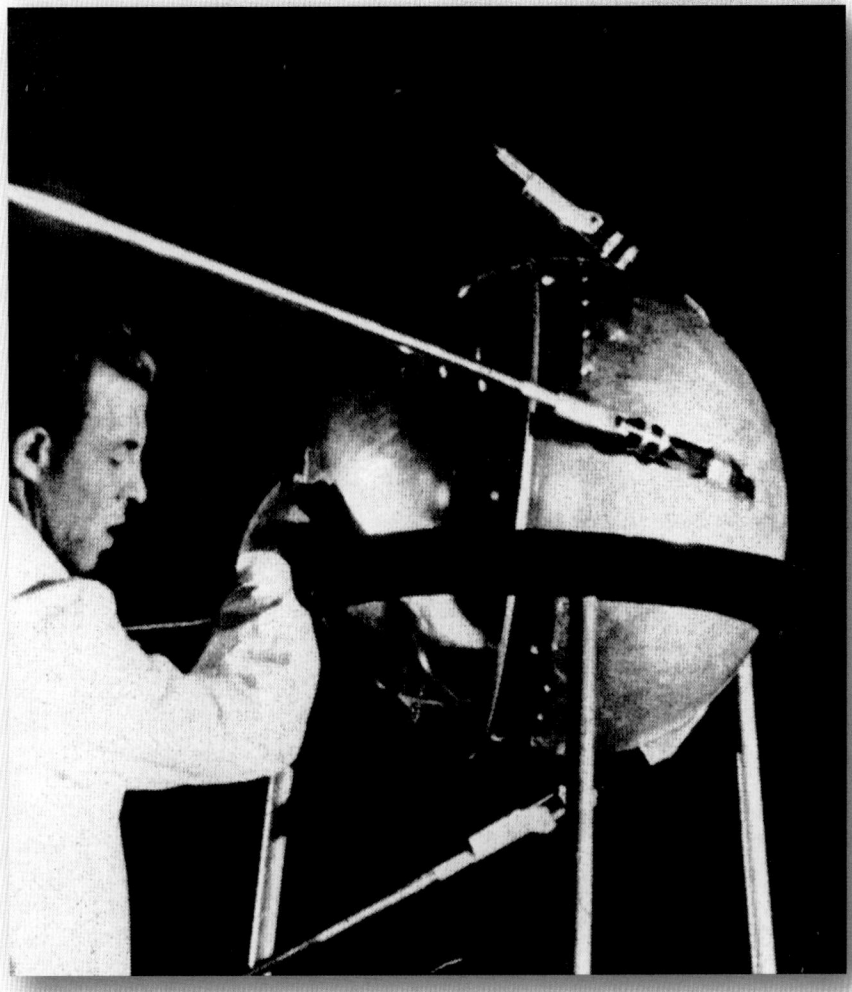

Sputnik 1

◀ Sputnik 1 (PS-1) was launched on October 4, 1957. It was the first human-made object in space. Developed in November 1956, it was launched when the original ambitious plan for a more complex satellite was shelved in favor of beating the Americans to get the first object in space.

The pressurized sphere—named Sputnik, or Satellite—was made of aluminum alloy and designed to test the method of placing an artificial satellite into Earth orbit, to provide information on the density of the atmosphere, to test radio and **OPTICAL** methods of orbital tracking, to examine the effects of **RADIO WAVES** moving through the atmosphere, and to check principles of pressurization used on the satellites.

To do this, it used temperature **SENSORS** pressurized with nitrogen gas. Changes in temperature caused changes in the signals sent to Earth. At the same time, the nature of the radio signals as they traveled to the Earth gave important information about the **IONOSPHERE**.

Sputnik was powered by internal batteries, and so its signal stopped when the batteries ran out 21 days after launch.

The satellite was tracked by observatories around the world, forming the first stage of the now vitally important **GLOBAL POSITIONING SYSTEM** (GPS) that is used for all kinds of modern surveying and direction finding.

The position of the satellite was found by its **DOPPLER EFFECT**, caused by its movement. As the satellite sped away from the observing station, its signal **FREQUENCY** changed in one direction, while as it moved toward another station, its signals appeared to change in the opposite direction.

On November 3, 1957, Sputnik 2 was put into orbit before a now awestruck world. It weighed 508.3 kg and carried the first living creature—a dog named Laika—into space.

Find about more about the global positioning system in Volume 8: *What satellites see*.

23

This space mission not only discovered many things about how living things survive in space, but also about SOLAR RADIATION and other RADIATION in space. For instance, it was learned that there was a radiation belt around the Earth.

The American challenge

Sputniks 1 and 2 represented one of human beings' greatest achievements: sending something out to beyond the Earth's ATMOSPHERE. Their effect was to electrify the world and to galvanize many countries into ever greater action—the main competitor for the Soviet Union being the United States.

Although the Soviet Union was the first to put something in space, the United States had also been developing space vehicles for decades. Less than 4 months after Sputnik the U.S. tried to launch Explorer, a mission speeded up by political pressure. The satellite to be carried was tiny: just 1.4 kg, no bigger than a grapefruit. But the small rocket pressed into use for this attempt could not even lift itself from the launch pad by more than a few meters.

It was clearly necessary to use a bigger launch vehicle, and that was provided by the military rocket named Jupiter.

Explorer and Vanguard

It was not until January 31, 1958, that the first successful launch of an American satellite occurred. The satellite weighed just over a kilogram and looked like a cylinder. It was called Explorer 1.

The Explorer series was the first successful step in the American challenge. Each satellite in the series was designed to find out more about the outer atmosphere where manned spacecraft would be flying. Explorer 1 carried GEIGER TUBES that detected radiation. The experiments were conducted by Dr. James Van Allen. The radiation belt around the Earth detected by these tubes would be called the Van Allen belt.

▲ The explosion of an early Navy Vanguard rocket. On December 6, 1957, carrying a satellite payload, it failed to develop enough THRUST and toppled over on the launch pad. So soon after the successful launch into orbit of Sputnik, Vanguard was nicknamed "kaputnik." It simply showed how difficult space launches are.

ARRAY A regular group or arrangement.

ATMOSPHERE The envelope of gases that surrounds the Earth and other bodies in the universe.

GEIGER TUBE A device to detect radioactive materials.

RADIATION The transfer of energy in the form of waves (such as light and heat) or particles (such as from radioactive decay of a material).

SOLAR CELL A photoelectric device that converts the energy from the Sun (solar radiation) into electrical energy.

SOLAR RADIATION The light and heat energy sent into space from the Sun.

THRUST A very strong and continued pressure.

THE SPACE RACE

◀ Explorer 6, with its "paddle-wheel" **ARRAYS** of **SOLAR CELLS**, was the first to take a TV picture of the Earth from space. Although it looks crude, it was an important step toward the high imagery standards that we now take for granted.

▼ The Explorer 17 satellite weighed 184 kg. It was an 89 cm stainless steel sphere designed to measure the density, composition, pressure, and temperature of Earth's atmosphere. It was launched on April 3, 1963.

▲ The Vanguard satellite launched on March 17, 1958, was the first to carry an alternative power source to batteries. It had solar cells. The solar cells were used to power the transmitter. Vanguard I had such a high orbit that it still has not reentered the Earth's atmosphere and is the oldest object left in space.

The Vanguard series was launched in an effort to find out about the Earth's **MAGNETIC FIELD** and to test the working of **SOLAR CELLS** (at that time quite revolutionary).

After the launch of the Sputniks 1, 2, and 3 the Soviet Union did not officially send up any more satellites (although its military program actually put up many satellites). Instead, its public efforts went into manned spaceflight. By contrast, the United States launched satellite after satellite in its attempts to master rocket and satellite technology.

Discoverer

The first of the Discoverer satellites was launched on February 28, 1959. Discoverer 13 contained a **CAPSULE** with a **HEAT SHIELD**, making it the first item to be returned safely to the Earth's surface. It landed by parachute in the sea. As people found out later, the Discoverer program was probably also the very first military spy satellite series, each containing high-resolution cameras.

▶ Discoverer 13 was the first space vehicle to be recovered from Earth orbit.

THE SPACE RACE

Tiros

Tiros (television and infrared observation satellite) was launched on April 1, 1960. It was the first weather satellite and transformed weather forecasting worldwide. Previously, forecasters had only been able to get records from the ground: They had no global view. Suddenly, they had pictures of all the cloud patterns of the world's weather system.

Thanks to Tiros satellite pictures of weather became commonplace on TV screens, so that people could see the approaching weather systems. Further applications included forecasting the weather for aircraft, ships, and farmers. The commercial application of the data from this type of satellite was one of the fastest on record.

▲ Tiros, the first weather satellite.

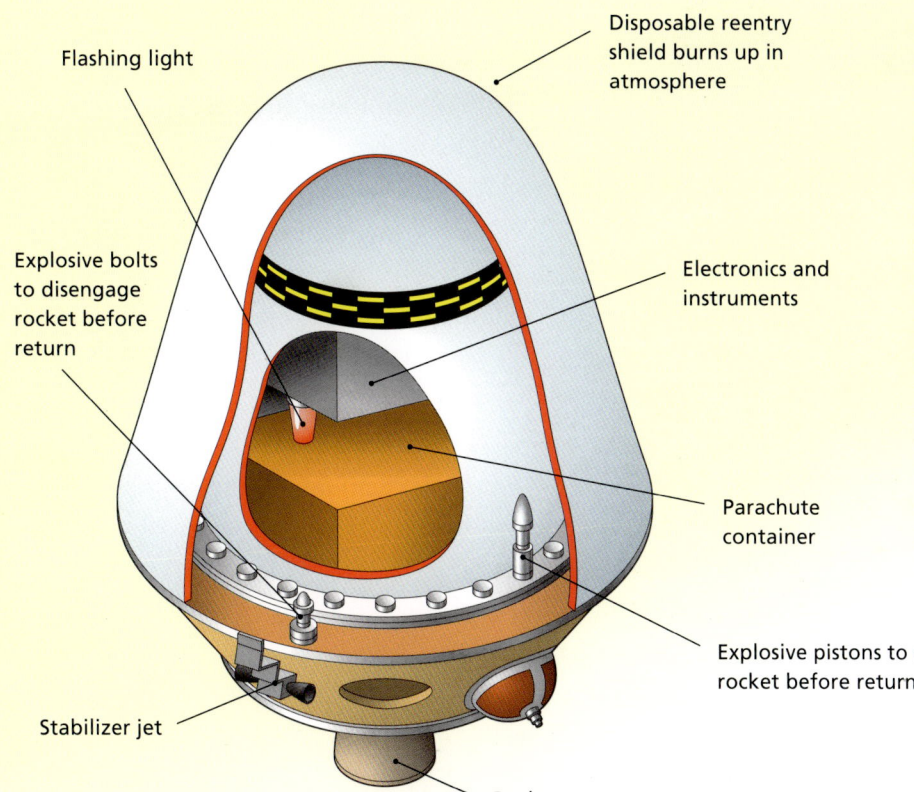

CAPSULE A small pressurized space vehicle.

HEAT SHIELD A protective device on the outside of a space vehicle that absorbs the heat during reentry and protects it from burning up.

MAGNETIC FIELD The region of influence of a magnetic body.

SOLAR CELL A photoelectric device that converts the energy from the Sun (solar radiation) into electrical energy.

27

By now it was standard to have solar cells recharging internal batteries so that satellites could function for years. Satellites also had data-storing facilities to keep data on tape and then transmit it when their orbit took them to within range of a **GROUND STATION**.

Telstar

Another important development was the launching of communications satellites.

It began with the Echo series of satellites launched in 1960. They were balloonlike, made of plastic with an aluminum coating. The balloon was inflated in space by using a substance called acetamide that turned from a solid to a gas (it sublimed). This balloon was 30 m across and could be seen clearly from

▲ The Echo series, the forerunner in communications technology from space.

the Earth's surface, looking like a slowly moving planet or star. Signals were bounced from it.

The first, and possibly the most famous satellite other than Sputnik, was Telstar. Inside Telstar was a very powerful amplifier so that it could pick up signals and magnify and retransmit them, rather than just bounce signals like the Echo satellites. Telstar was so revolutionary that it even had a hit pop song written for it. It was launched in 1962. The purpose of the satellite was to receive and store messages and retransmit them on demand.

The dramatic innovation that Telstar brought was the transmission of the first live pictures across the Atlantic. Using Telstar, a transmitter in the UK was able to beam the first live television pictures to America. On July 12, 16 European stations exchanged television programs with the United States.

Telstar led the way in the development of the global orbiting **GEOSTATIONARY SATELLITES**, which we all now use for our international (and some of our national) communications.

THE SPACE RACE

▼ Telstar, the satellite that was to lead a communications revolution. It was almost a meter across and weighed 80 kg. It was powered by 3,600 solar cells and used a new traveling-wave radio tube as its microwave radio source.

The satellite had a capacity of 600 voice channels or one television channel. Telstar 1 was in service for only about a year, but was followed by a long succession of increasingly sophisticated satellites that also carried the Telstar name.

GEOSTATIONARY SATELLITE A man-made satellite in a fixed or geosynchronous orbit around the Earth.

GROUND STATION A receiving and transmitting station in direct communication with satellites. Such stations are characterized by having large dish-shaped antennae.

For more on geostationary satellites and orbits see Volume 8: *What satellites see*.

4: MANNED SPACEFLIGHT

In just 5 years the world had gone from having no **SATELLITES** to having many in space performing a wide range of tasks. It was one of the most staggering developments in technology the world has ever known.

It was now possible to launch satellites, put them in **ORBIT**, and use them. It was even possible to recover **CAPSULES** from them. But none of this would allow people to survive in space.

Looking back 5 years, the Soviet Union had launched a dog into orbit, but since then there appeared to have been no further steps for manned spaceflight. But behind the scenes both the Soviet Union (through their Vostok trials) and the United States (through their Mercury trials) were working furiously to find ways to put people in space. Through these tests, using both dogs and chimpanzees, both countries were learning more and more about how to sustain life in space.

CAPSULE A small pressurized space vehicle.

GROUND STATION A receiving and transmitting station in direct communication with satellites. Such stations are characterized by having large dish-shaped antennae.

ORBIT The path followed by one object as it tracks around another.

SATELLITE A man-made object that orbits the Earth.

▶ Mercury capsule in final assembly before launch.

MANNED SPACEFLIGHT

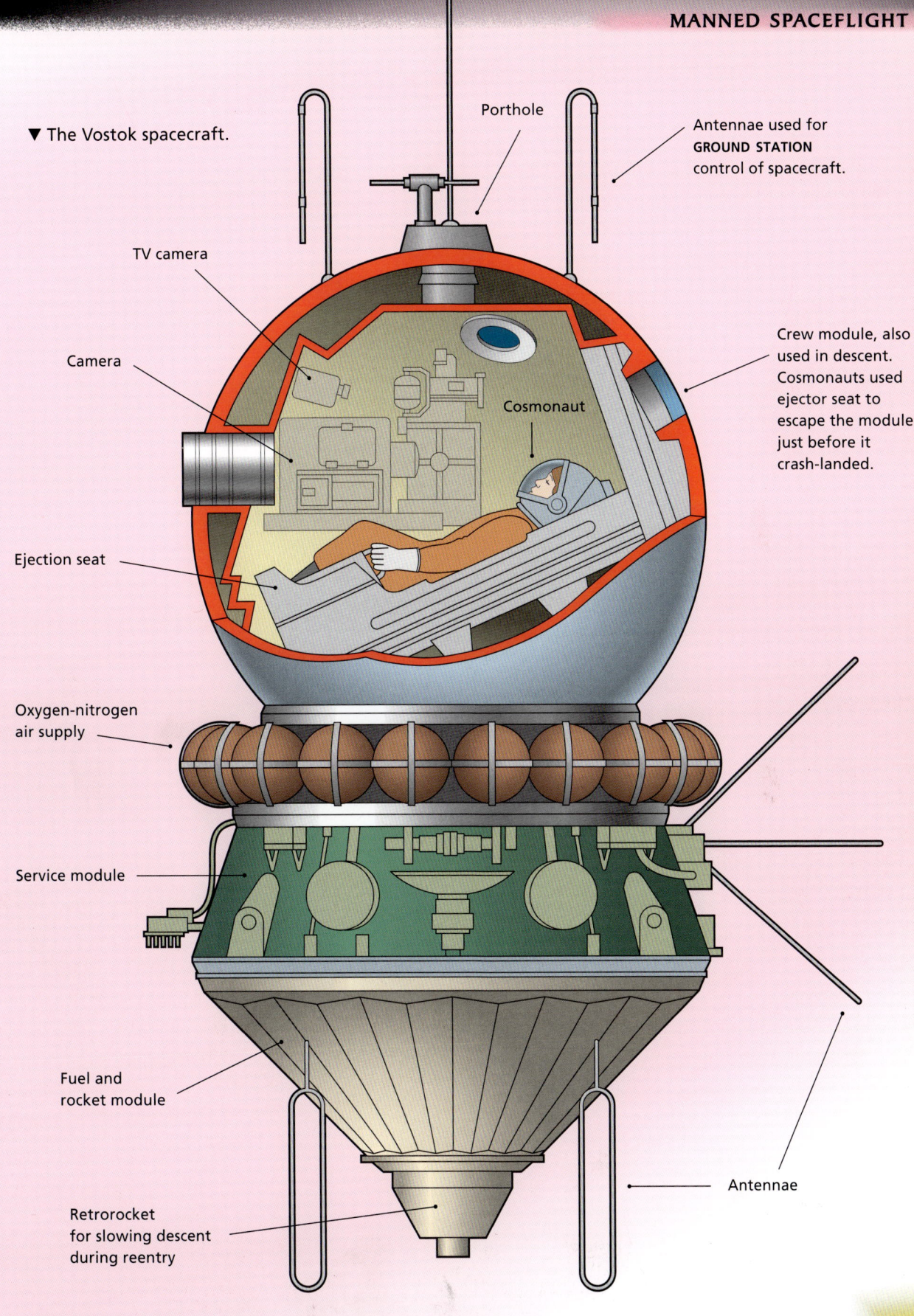

▼ The Vostok spacecraft.

MANNED SPACEFLIGHT

GRAVITY The force of attraction between bodies. The larger an object, the more its gravitational pull on other objects.

HEAT SHIELD A protective device on the outside of a space vehicle that absorbs the heat during reentry and protects it from burning up.

RETROROCKET A rocket that fires against the direction of travel in order to slow down a space vehicle.

Vostok

The Soviet Union was again first to launch a man into orbit around the Earth. Yuri Gagarin will be one of the most remembered people in history for his epic journey on April 12, 1961.

The spacecraft that the Soviet Union used was so primitive that Gagarin had no control over it at all—he was more like a passenger than a pilot. His Vostok spacecraft was nearly 4.5 m long. The reentry capsule was a sphere just bigger than the height of the pilot, a mere 2.3 m across.

The reentry vehicle had no way to land, even in the sea, and so Gagarin had to press an ejector button to fire himself clear of the reentry vehicle so that he could land by parachute, while the vehicle crashed to the ground.

The **HEAT SHIELD** of the Vostok was designed to burn away during reentry.

Inside the capsule Gagarin had food, a radio, and a porthole to see through. He also had a box containing scientific equipment for experiments.

Attached to the capsule was an instrument section containing the **RETROROCKET** that was needed to slow the capsule during reentry.

It was important to play safe, because much of the equipment was still untested or unreliable. For example, Gagarin was launched into a low orbit so that he would return by **GRAVITY** if the retrorocket failed to work.

◀ Yuri Gagarin in the bus on the way to the launch pad on the morning of April 12, 1961. Behind him, seated, is his backup, German Titov.

On April 12, 1961, at 9:06 A.M. Gagarin lifted off in the Vostok 1 spacecraft. After a 108-minute flight of extended low gravity he parachuted safely to the ground in the Saratov region of the USSR. The first human to fly in space, he successfully completed one orbit around the Earth. After his historic flight Gagarin became an international symbol for the Soviet space program and in 1963 was appointed deputy director of the Cosmonaut Training Center. In 1966 he served as a backup crew member for Soyuz 1.

33

▲ Alan Shepard inside the Mercury 3 mission capsule, Freedom 7.

Mercury

The American initiative was named Mercury. The first flight in the series was not an orbital flight but a simple arching path, or **TRAJECTORY**. It was performed on May 5, 1961, about 3 weeks after the Vostok launch.

Alan B. Shepard, Jr., was the first person to fly under the Mercury program, housed in a 1,400 kg Mercury space **CAPSULE** called Freedom 7. It was powered by a Redstone rocket. The nearly 500-km journey lasted just 15 minutes.

▼ Astronaut John Glenn entering the Mercury 6 mission capsule, Friendship 7, and about to begin the first American manned Earth orbital mission.

CAPSULE A small pressurized space vehicle.

TRAJECTORY The curved path followed by a projectile.

MANNED SPACEFLIGHT

▶ Astronaut Walter M. Schirra, Jr., in Mercury pressure suit with a model of the Mercury capsule behind him.

Although his flight was less spectacular than the Vostok orbit, there was an important difference. Shepard was able to help control his capsule and thus land in it on the sea. As a result, although superficially it was not yet apparent, the race was becoming more even.

The Redstone rocket got the program under way, but it was not powerful enough for an orbital launch. The succeeding launches used the more powerful Atlas rockets.

When John Glenn piloted the Mercury 6 capsule Friendship 7 on February 20, 1962, he made three orbits. Although still 14 orbits behind the Soviet cosmonauts' record at that stage, Glenn's mission was an American landmark, and he became a national hero.

NASA then began a series of further orbiting runs, with Wally Schirra doing nine orbits in Sigma 7 and Gordon Cooper, Jr., performing 22 orbits on May 15, 1963, with a space duration of 34 hours and 20 minutes.

Not to be put in the shade, in June 1963 the Soviets launched Valentina Tereshkova—the first woman into space. While in orbit, she passed another cosmonaut, Valeri Bykovsky, who in his 5-day journey was performing the longest solo spaceflight in history so far.

▲ Three women cosmonauts prior to the launch of Vostok 6, June 1963. Left to right are Valentina Ponomareva, backup Irina Solovyeva, and Valentina Tereshkova—the first woman in space. Tereshkova spent 3 days aboard Vostok 6.

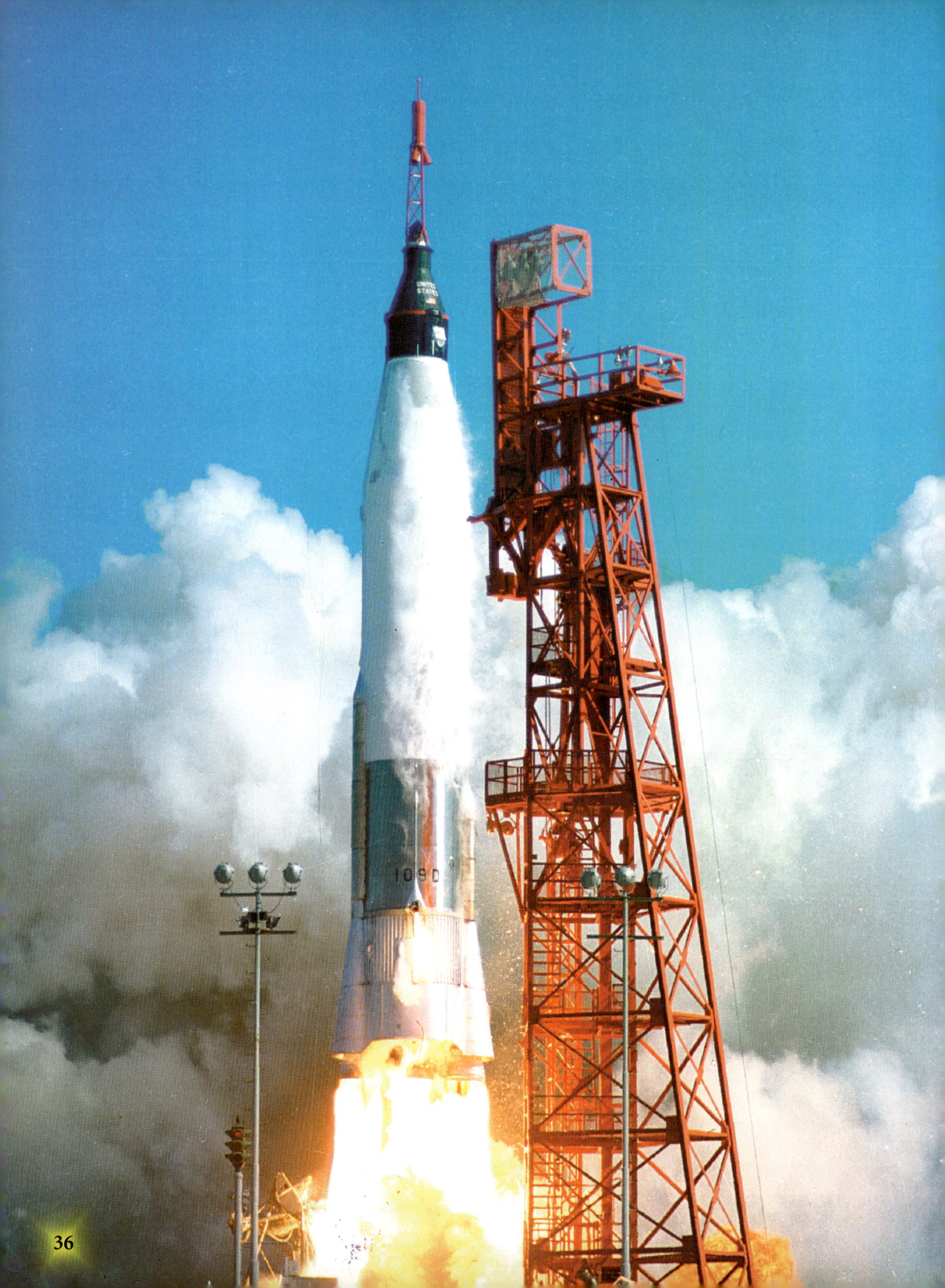

MANNED SPACEFLIGHT

◀ Friendship 7 takeoff.

▲ Navy divers install a stabilizing flotation collar around Gordon Cooper's Mercury space capsule, nicknamed "Faith 7."

5: TO THE MOON

In May 1961 President John F. Kennedy announced to an astonished world that by 1969 America would land people on the Moon. To achieve this, astronauts and scientists would have to find out the effects on people of long flights, how to maneuver a spacecraft, how to meet and **DOCK** with another spacecraft, and how to walk in space (now called EVA, or **EXTRAVEHICULAR ACTIVITY**).

The program to accomplish this bold move to the boundaries of science and technology was to be called Gemini, and it would be a triumph of using (simple) computer technology. It was also a feat that the Soviet Union did not have the resources to match.

Gemini

Gemini was a two-astronaut, bell-shaped spacecraft made up of the reentry **CAPSULE** and a section containing the **RETROROCKETS**. There was also an on-board computer.

To get these spacecraft into **ORBIT**, a new **LAUNCHER** was needed. It was called Titan 2.

There were ten manned spacecraft missions, each designed to test yet one more aspect of the techniques needed to accomplish the journey to the Moon.

By Gemini 4 the first American was making an EVA. The combination of Gemini 6 and 7 produced the first rendezvous in space. They did not dock, but came within 30 cm of one another.

▶ Gemini liftoff using the Titan 2 launcher.

CAPSULE A small pressurized space vehicle.

DOCK To meet with and attach to another space vehicle.

EXTRAVEHICULAR ACTIVITY Any task performed by people outside the protected environment of a space vehicle's pressurized compartments. Extravehicular activities (EVA) include repairing equipment in the Space Shuttle bay.

LAUNCHER A system of propellant tanks and rocket motors or engines designed to lift a payload into space. It may, or may not, be part of a space vehicle.

ORBIT The path followed by one object as it tracks around another.

RETROROCKET A rocket that fires against the direction of travel in order to slow down a space vehicle.

▶ Spectacular EVA on Gemini 4.

TO THE MOON

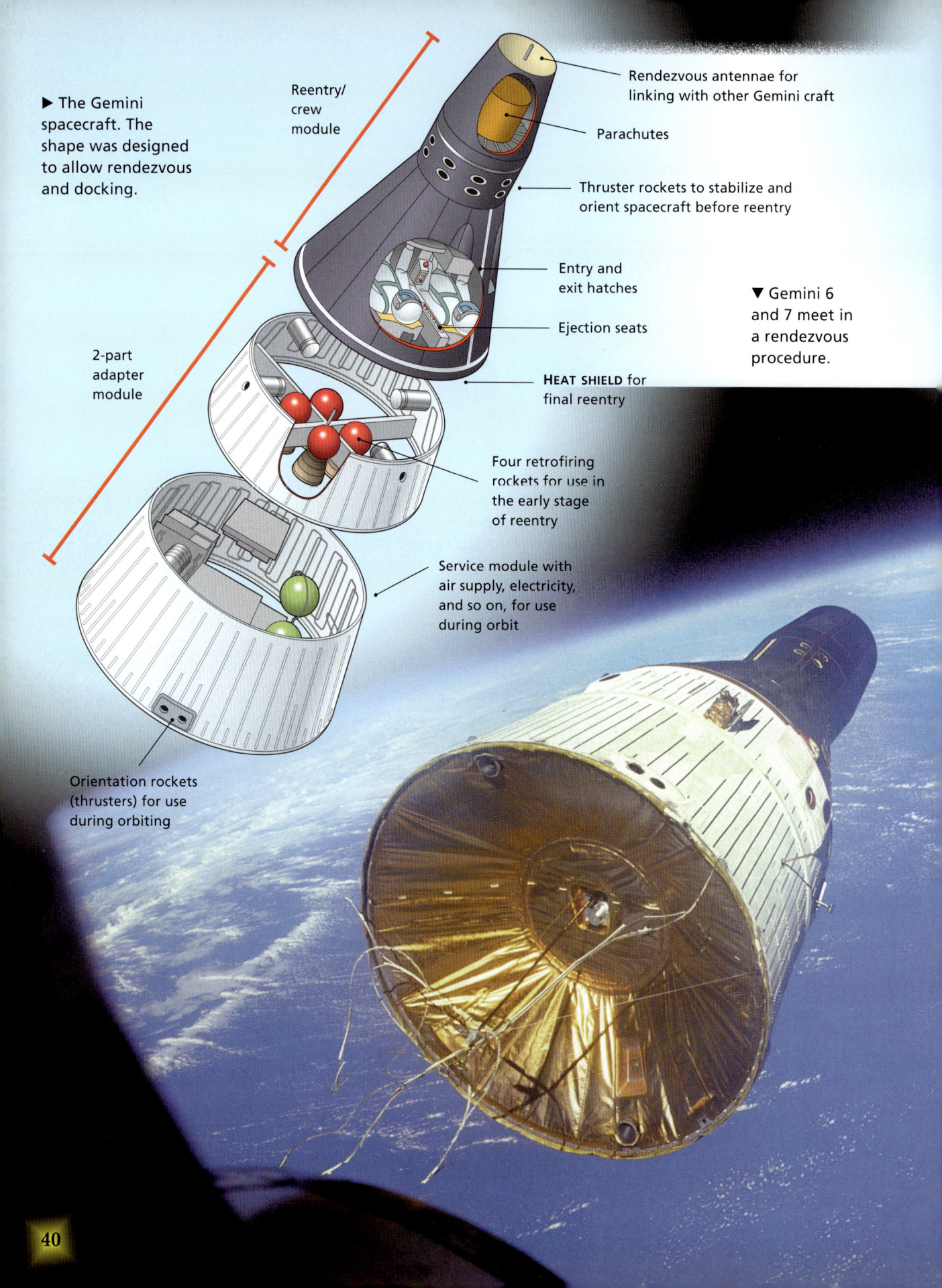

▶ The Gemini spacecraft. The shape was designed to allow rendezvous and docking.

Reentry/crew module

2-part adapter module

Rendezvous antennae for linking with other Gemini craft

Parachutes

Thruster rockets to stabilize and orient spacecraft before reentry

Entry and exit hatches

Ejection seats

HEAT SHIELD for final reentry

Four retrofiring rockets for use in the early stage of reentry

Service module with air supply, electricity, and so on, for use during orbit

Orientation rockets (thrusters) for use during orbiting

▼ Gemini 6 and 7 meet in a rendezvous procedure.

TO THE MOON

▶ The recovery phase after the landing of Gemini 7. This picture gives a clear insight into the amount of space each astronaut had, together with their positions in the capsule. Notice also the large hatches for ease of entry and exit, both on the ground and in space during an EVA.

The actual first docking was by Gemini 8, but it had problems, as did Gemini 9. During the space walk of Gemini 9, in which Gene Cernan tried to detach a part that had got stuck, astronauts had their first experience of **GRAVITY** not being there to help them push against an object. As Newton's **LAWS OF MOTION** state, an action is met by an equal and opposite **REACTION**. As Cernan went to touch something, the reaction force bounced him away from it.

It was Gemini 10 that actually docked first, using an Agena rocket. The rocket was then used to boost the altitude of Gemini into a new orbit, a maneuver crucial to further developments in space travel.

GRAVITY The force of attraction between bodies.

HEAT SHIELD A protective device on the outside of a space vehicle that absorbs the heat during reentry and protects it from burning up.

LAWS OF MOTION Formulated by Sir Isaac Newton, they describe the forces that act on a moving object.

REACTION An opposition to a force.

41

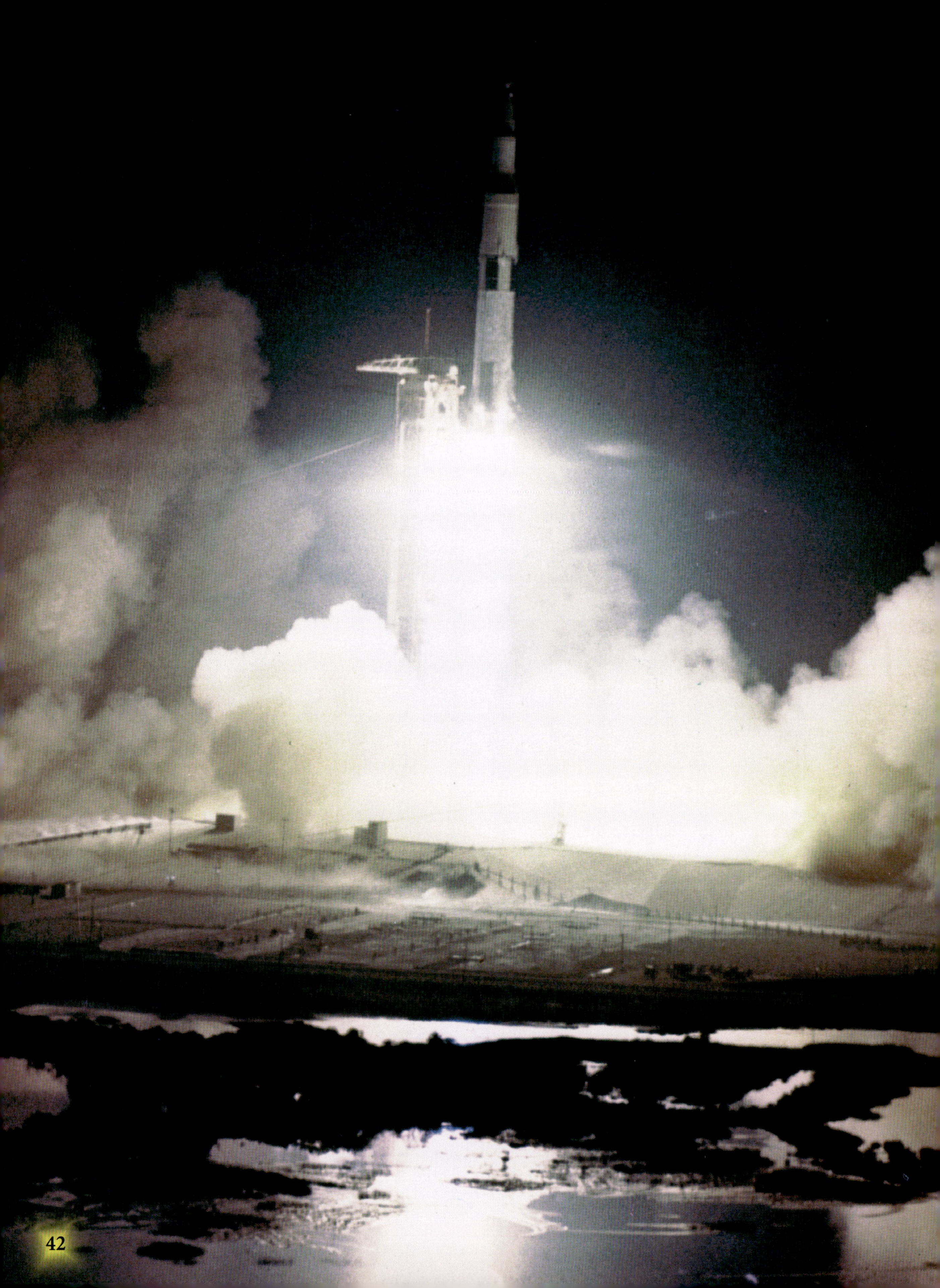

TO THE MOON

PROPULSION SYSTEM The motors or rockets and their tanks designed to give a launcher or space vehicle the thrust it needs.

TRAJECTORY The curved path followed by a projectile.

The Apollo project

The project to reach the Moon was called Apollo (after Apollo, the god of light).

It involved a very powerful launcher called Saturn V (see page 20). It was needed because the spacecraft it was to launch into a lunar **TRAJECTORY** weighed 50 tonnes.

The Apollo spacecraft, consisting of the command module (CM) and service module (SM) (together known as the CSM), as well as the lunar module (LM), had to have sufficient rocket power to land and take off from the Moon's surface as well as to send itself into another suitable trajectory for reentering the Earth's atmosphere. To save the extra power required in getting a large spacecraft off the Moon's surface, only the lunar module was used in the descent to the Moon. It was nicknamed "Eagle." While the LM descended to the Moon, the CSM continued to orbit the Moon.

The command module was the spacecraft's control center. It was designed to be big enough to house three astronauts. It was cone-shaped, 3.7 meters high, 4 meters in diameter, and weighed about 4,500 kilograms.

The service module was fitted below the command module. It was 3.9 meters in diameter and 6.7 meters long. Its purpose was to house the **PROPULSION SYSTEM** that would make any course corrections, would retrofire rockets to put the spacecraft into lunar orbit, and later would send the spacecraft out of lunar orbit and into a path that would return it to the Earth.

The lunar module was positioned between the service module and the Saturn V launch vehicle. It was the part of the spacecraft that would carry two astronauts on the short journey from the lunar orbit to the surface of the Moon. When their surface explorations were over, it would then return them to the command module. The lunar module was about 6.4 meters high and 3.4 meters in diameter, and weighed 13,600 kilograms.

◀ Apollo takeoff.

1
Leaving Earth

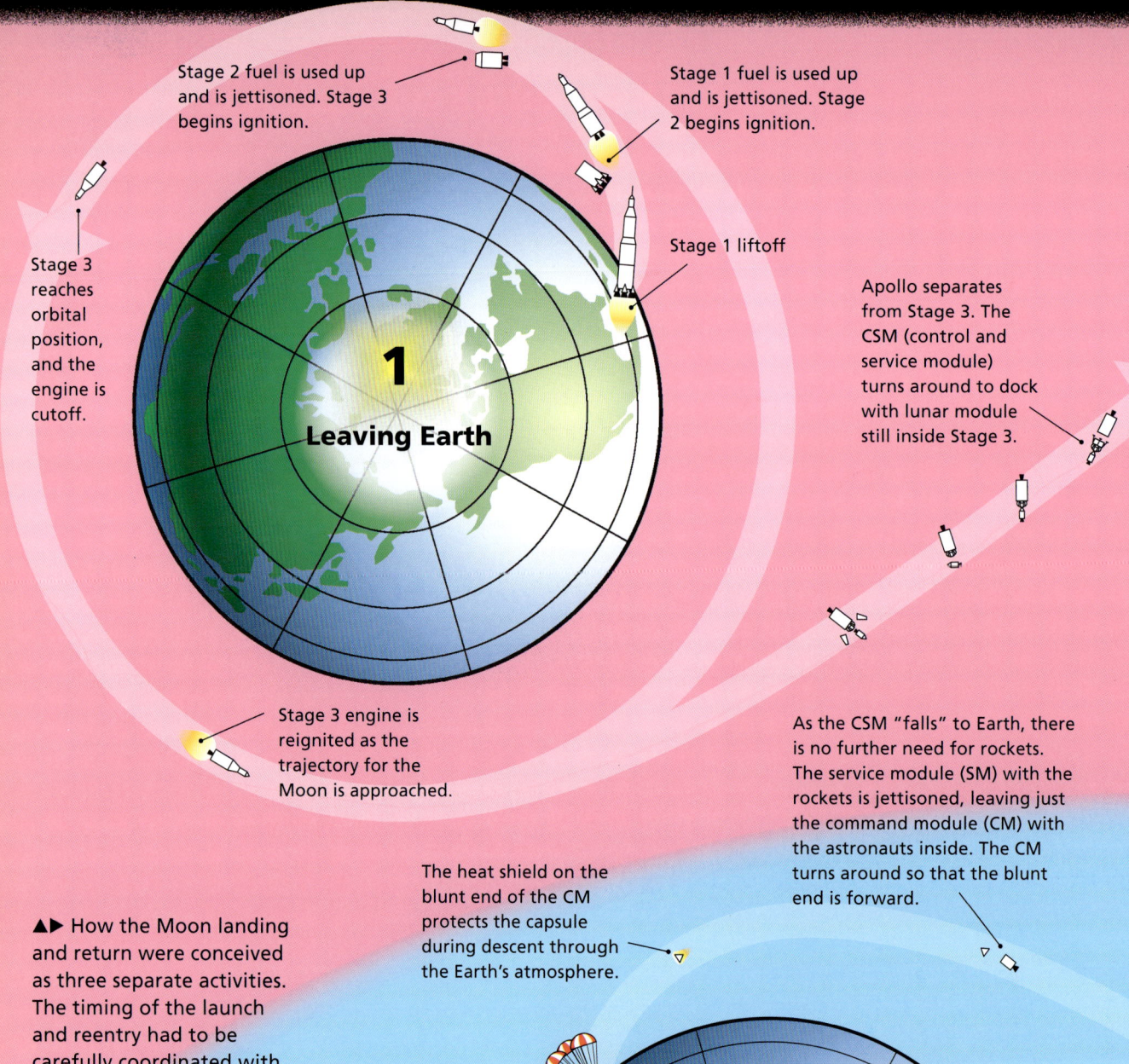

Stage 2 fuel is used up and is jettisoned. Stage 3 begins ignition.

Stage 1 fuel is used up and is jettisoned. Stage 2 begins ignition.

Stage 1 liftoff

Stage 3 reaches orbital position, and the engine is cutoff.

Apollo separates from Stage 3. The CSM (control and service module) turns around to dock with lunar module still inside Stage 3.

Stage 3 engine is reignited as the trajectory for the Moon is approached.

As the CSM "falls" to Earth, there is no further need for rockets. The service module (SM) with the rockets is jettisoned, leaving just the command module (CM) with the astronauts inside. The CM turns around so that the blunt end is forward.

The heat shield on the blunt end of the CM protects the capsule during descent through the Earth's atmosphere.

▲▶ How the Moon landing and return were conceived as three separate activities. The timing of the launch and reentry had to be carefully coordinated with the changing position of the Moon with respect to the Earth. The diagram is not to scale.

Parachute is used for splashdown in the Pacific Ocean.

44

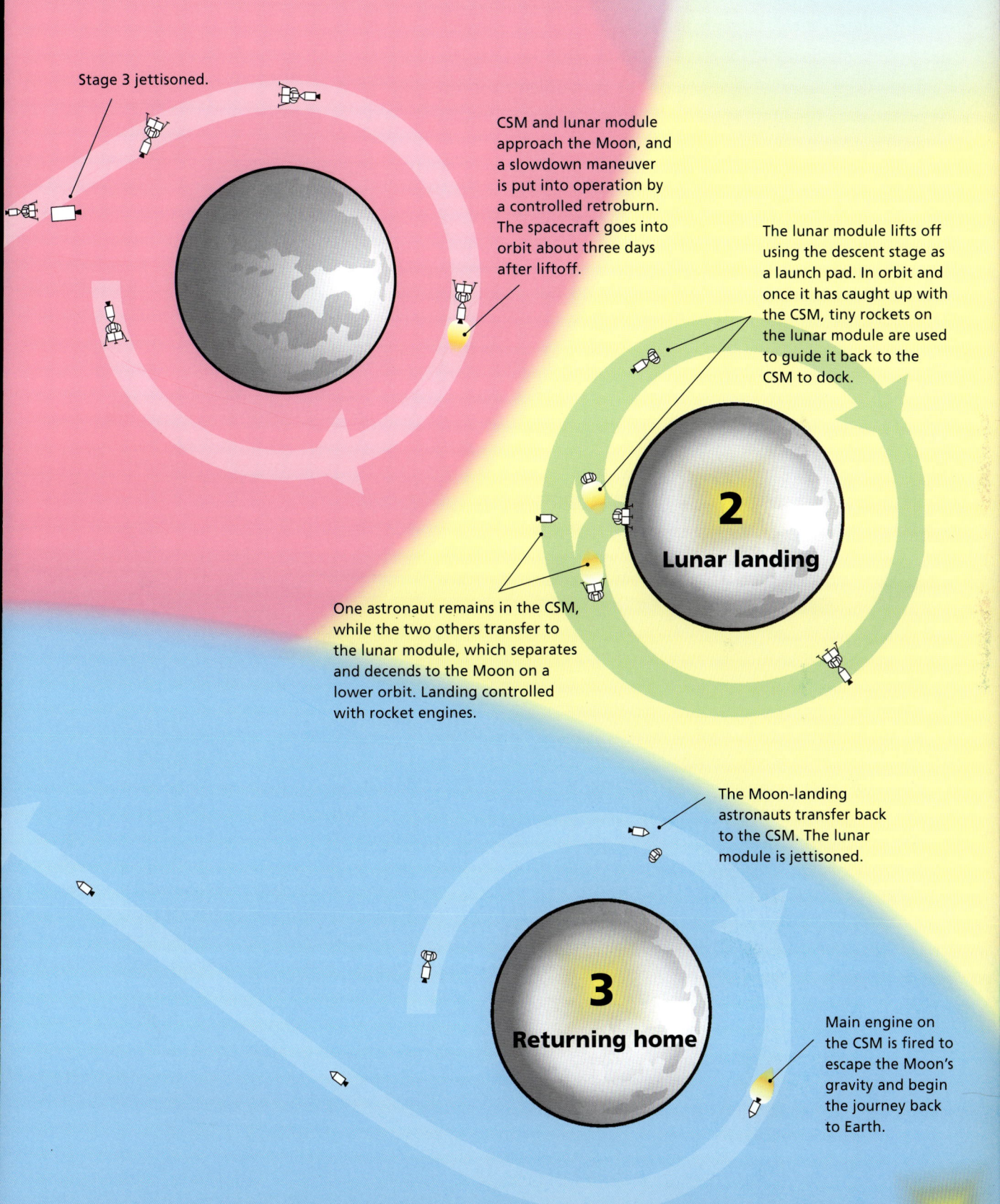

TO THE MOON

Stage 3 jettisoned.

CSM and lunar module approach the Moon, and a slowdown maneuver is put into operation by a controlled retroburn. The spacecraft goes into orbit about three days after liftoff.

The lunar module lifts off using the descent stage as a launch pad. In orbit and once it has caught up with the CSM, tiny rockets on the lunar module are used to guide it back to the CSM to dock.

One astronaut remains in the CSM, while the two others transfer to the lunar module, which separates and decends to the Moon on a lower orbit. Landing controlled with rocket engines.

2 Lunar landing

The Moon-landing astronauts transfer back to the CSM. The lunar module is jettisoned.

3 Returning home

Main engine on the CSM is fired to escape the Moon's gravity and begin the journey back to Earth.

45

During the flight between the Earth and the Moon the CSM would have to be decoupled from the Stage 3 rocket and turned around so that it could DOCK with the LM. Only then could the assembled spacecraft be decoupled from the Stage 3 rocket. The final spacecraft traveled with the lunar module in front.

The lunar module housed two engines. One of them could be varied in THRUST to allow controlled descent to the surface of the Moon. The other was for ascent back to the command module. The lunar module itself was made of two parts. One part was used for descent, and the other part used the descent module as a launch pad to get back into lunar orbit. This was important since nobody knew what the surface of the Moon was like, and it was essential to have a stable platform from which to lift off.

▼ The command module was where the astronauts lived on their journeys to and from the Moon. It was also the part of the space vehicle that orbited the Moon while the lunar module, or Eagle, landed on the Moon's surface.

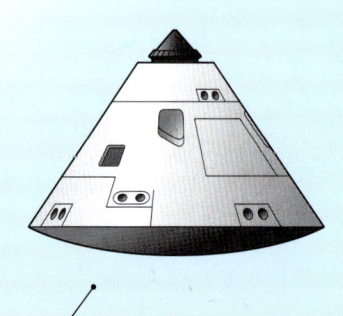

▼ Command/service modules assembled.

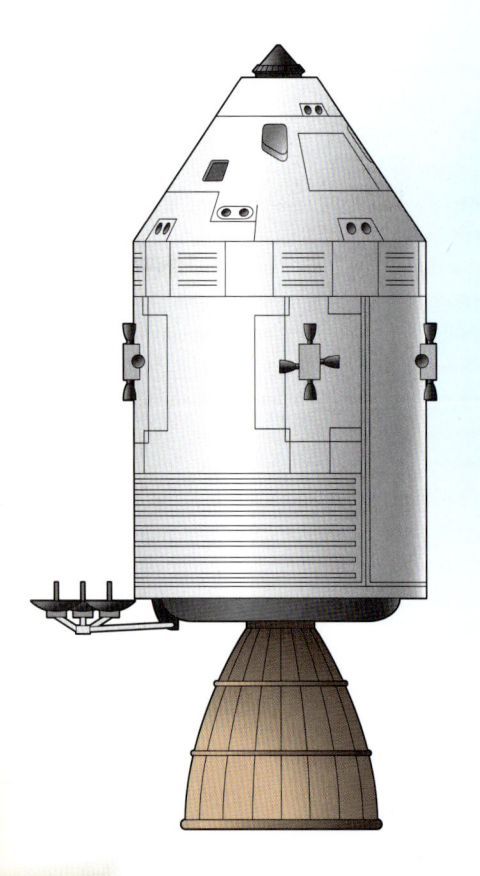

The command module reentered the atmosphere with this end leading. It was protected with a heat shield.

The service module provided the essential bulky services for the astronauts in the command module. It was attached to the command module for all of the journey except reentry. It housed the fuel for the rocket used to maneuver the spacecraft as well as oxygen tanks for the astronauts in the command module and the electricity supply to run the command module. The service module was jettisoned just before reentry.

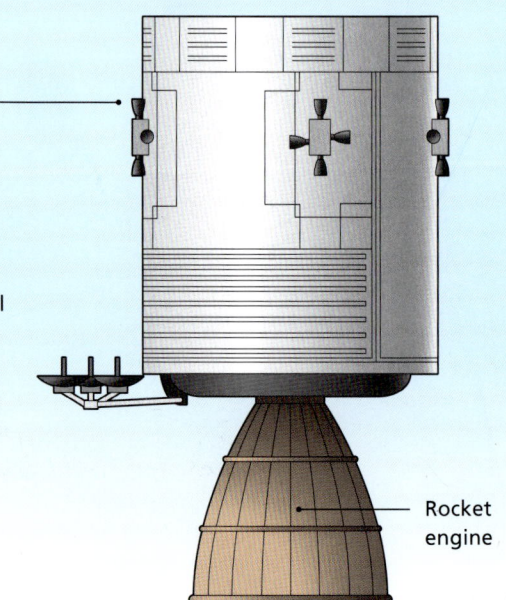

Rocket engine

46

TO THE MOON

To the Moon—carefully

Nobody wanted to embark into such uncharted territory without many small, careful preparatory steps. Several Apollo missions were completed without touching down on the Moon's surface. Apollo 7 made a 163-orbit flight carrying a full crew of three astronauts. Then Apollo 8 took a lunar **TRAJECTORY**, completed a lunar orbit, and returned safely to Earth.

Apollo 9 did not go to the Moon but spent time in Earth orbit testing the lunar module. Apollo 10 did go to the Moon but tested the lunar module without landing.

DOCK To meet with and attach to another space vehicle.

THRUST A very strong and continued pressure.

TRAJECTORY The curved path followed by a projectile.

▼ The complete lunar module, or Eagle.

This part of the lunar module housed two astronauts during touchdown and also contained the engine used for liftoff and maneuvering in order to rejoin the command and service modules.

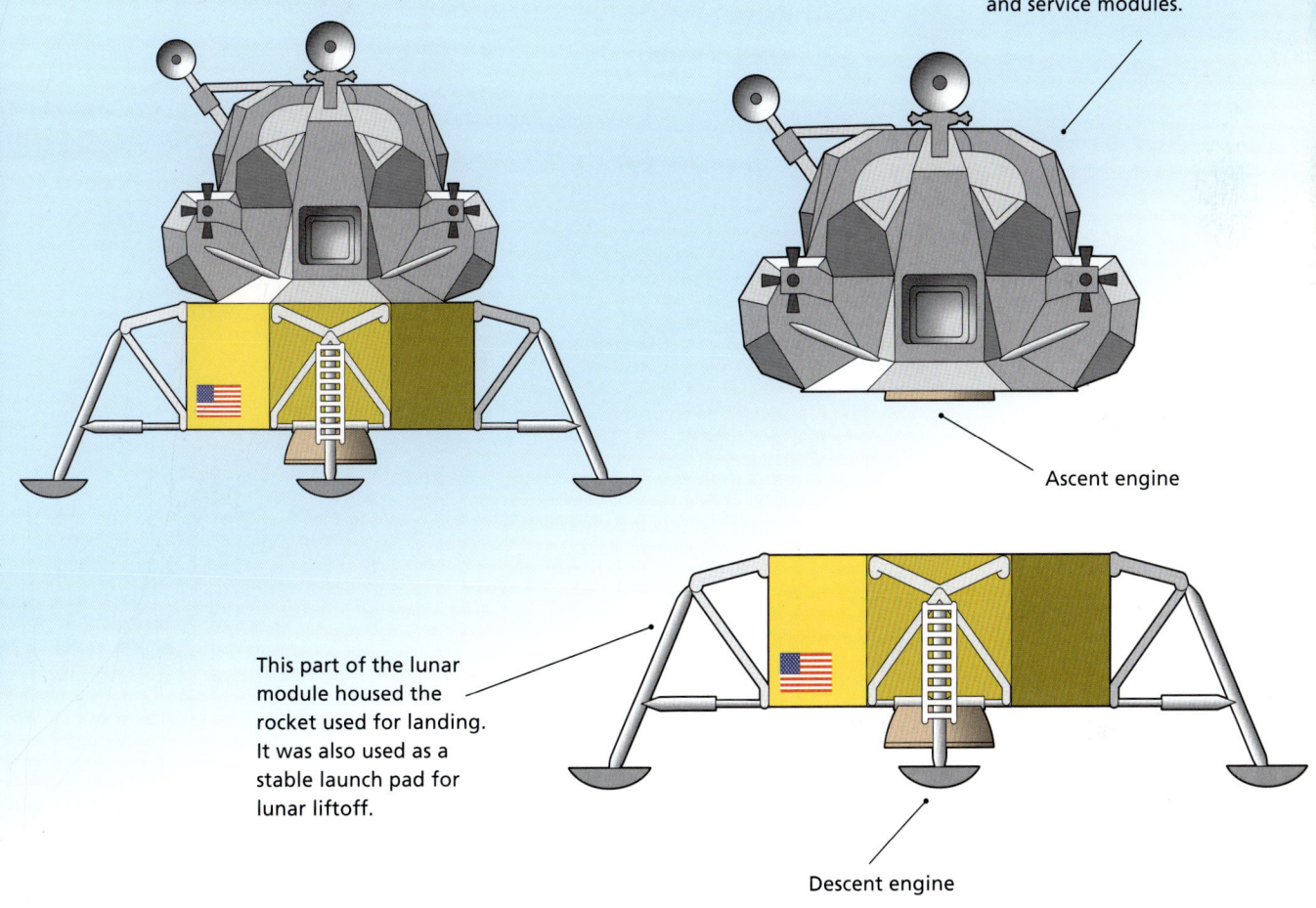

Ascent engine

This part of the lunar module housed the rocket used for landing. It was also used as a stable launch pad for lunar liftoff.

Descent engine

47

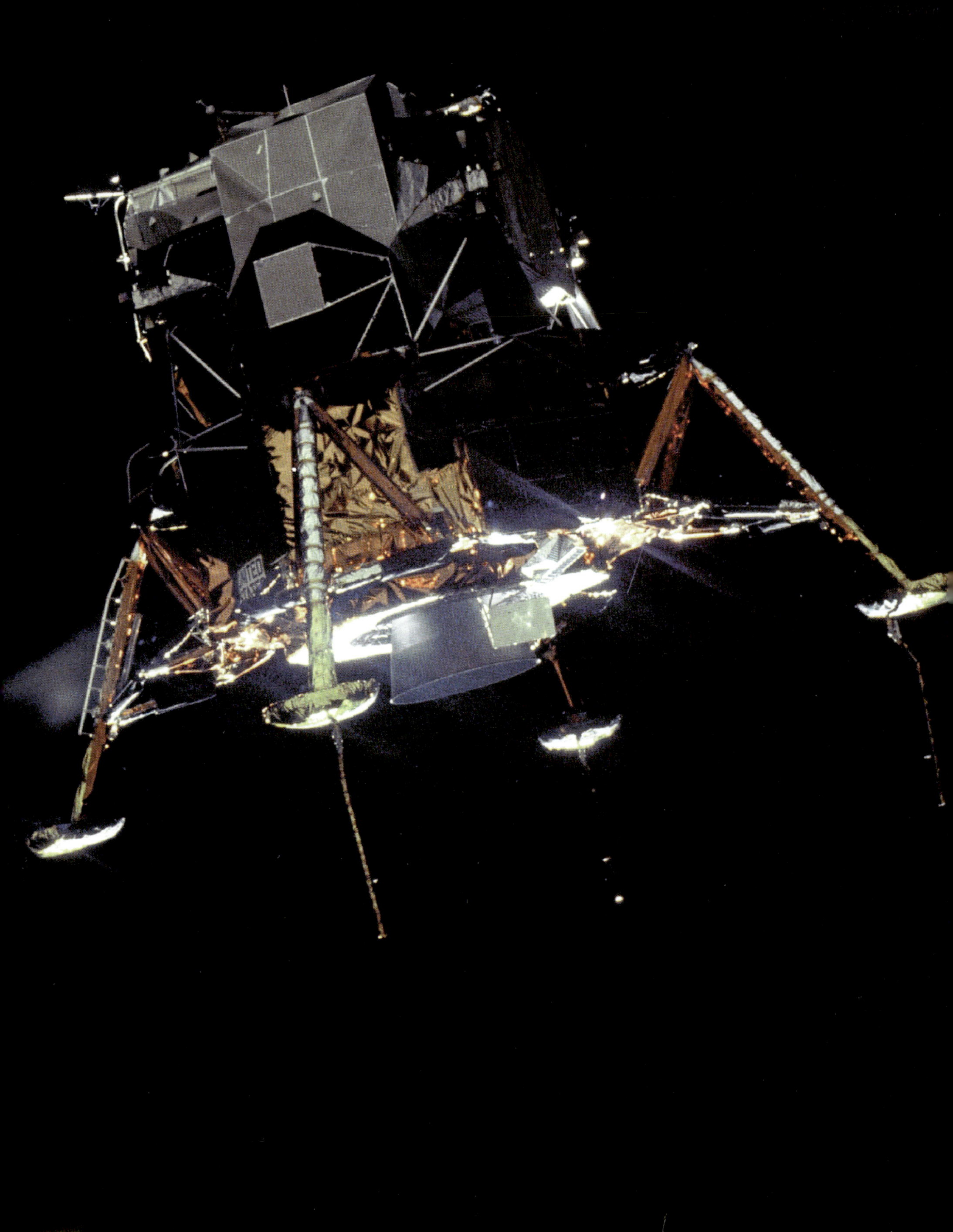

TO THE MOON

◀ The lunar module called Eagle. Notice the complete absence of any kind of streamlining because there is no atmosphere on the Moon and so no need to design for the effect of atmosphere.

The long rodlike protrusions under the landing pods are lunar surface sensing probes. When they touched the lunar surface, the probes sent a signal to the crew to shut down the descent engine.

Then finally, on July 20, 1969, Apollo 11's lunar module, with two astronauts on board, made a lunar landing—with astronaut Neil Armstrong being the first human to set foot on the Moon's surface.

As Neil Armstrong stepped onto the Moon's surface at 2:56 A.M. GMT on July 21, his first words were "That's one small step for man, one giant leap for mankind."

They remained on the surface for a day, setting up instruments and collecting samples before returning to the command module. After the LM and the CSM were docked, the astronauts had to return to the CSM. Then the LM could be decoupled and left in orbit while small rockets in the base of the CSM set it on a path that would eventually allow it to free fall to the Earth.

For more on the surface of the Moon see Volume 3: *Earth and Moon*.

▼ One of the first human footprints on the surface of another body in space.

▶ E."Buzz" Aldrin, Jr., lunar module pilot, seen during the historic first **EXTRAVEHICULAR ACTIVITY** on the Moon. He has just deployed the early Apollo scientific experiments package (EASEP). In the foreground is the passive seismic experiment package (PSEP); beyond it is the laser ranging retroreflector (LR-3); in the left background is the black-and-white lunar surface television camera; in the far right background is the lunar module Eagle. Astronaut Neil Armstrong, Apollo 11 mission commander, took this photograph.

▼ Walking on the Moon was "one small step for man, one giant leap for mankind." This picture shows Neil Armstrong at the modular equipment storage assembly (MESA) of the lunar module Eagle.

TO THE MOON

EXTRAVEHICULAR ACTIVITY Any task performed by people outside the protected environment of a space vehicle's pressurized compartments.

▲ The Moon rover enabled the astronauts on later Moon missions to explore widely from their landing module.

▲ Apollo 11 capsule recovered from the sea.

The entire journey from the Moon to the Earth thus relied on the attractive force of the Earth's gravity. No additional rocket thrust was needed. That allowed the CSM to jettison the SM and its rocket and to turn around (while still under free fall) so that the blunt heat shield end faced forward. The touchdown was in the Pacific Ocean on July 24, 1969.

Because it was not known if there was life on the Moon, the CM was carefully cleaned, and the astronauts were kept isolated.

Apollo 15, launched on July 26, 1971, carried a Moon rover, a four-wheeled, battery-powered lunar travel vehicle that allowed the astronauts to explore greater areas of the Moon's surface.

Between then and December 1972 more missions were completed, taking a total of 12 people to the Moon's surface. Finally, this hugely costly, but immensely satisfying, program was brought to a close.

6: OUTER WORLDS

To see the results from the space probes see Volumes 4, 5, and 8.

Manned spaceflights have told us much about how to survive in space and how to land on a planetary object. But we would still know next to nothing about the other planets in the solar system if manned spaceflight had been the only part of the space adventure.

Because of the enormous time involved in any journey and the difficulty of providing food, oxygen, water, and fuel, it is impossible at this time to send people to even the closest of the planets in the solar system. So, this journey of exploration has to be by unmanned space PROBE. Many probes have been sent to investigate other parts of the solar system, including some that have landed on other planets.

Reaching distant planets

If you want to reach a distant planet, it is not simply a matter of launching toward the visible planet in the sky. Every planet follows an orbit, many of which are highly elliptical (oval). The best time to reach such planets is when they are closest to the Earth. That reduces the amount of fuel that has to be used up in getting to them.

The TRAJECTORY needed to reach the planets has to be carefully calculated. If the LAUNCHER cannot provide enough VELOCITY for the space probe, its speed can be increased by aiming it close to another planet. This so-called SLINGSHOT TRAJECTORY allows the probe to come within the GRAVITATIONAL FIELD of a planet so that MOMENTUM can be transferred from the planet to the probe (see page 56). This system has been used for long-distance probes such as Mariner 10 to fly past Mercury and Pioneer 11 and Voyagers 1 and 2 to travel from Jupiter to Saturn.

Mariner 2
- Control system
- SOLAR PANELS
- TRUSSES connecting parts of the craft
- Communications dish (antenna)

Mariner
▲▼ The Mariner project was a series of space probes sent to Venus, Mars, and Mercury between 1962 and 1975. Later, Mariners obtained clear pictures of the Martian surface and were able to analyze the atmosphere. Mariner 10 gave the first closeup pictures of the surface of Mercury.

Mariner 5
- Control system
- Communications dish (antenna)
- Solar panels

52

OUTER WORLDS

Early probes

In 1962 Mariner 2 was the first spacecraft to fly past Venus. It had no cameras, but recorded temperature measurements and showed conclusively that Venus was extremely hot.

Apart from the antenna on the base of the spacecraft, necessary because of the long distance back to Earth, Mariner 2 was very similar to the Ranger series of spacecraft sent to photograph the Moon between July 1964 and March 1965.

The first Venus orbiters were Venera 9 and 10 in October, 1975, arriving on the 22nd and 25th, respectively. When they got there, they dropped the first landers to return pictures from the surface.

▼ The surface of Venus as photographed by Venera.

Parachute container

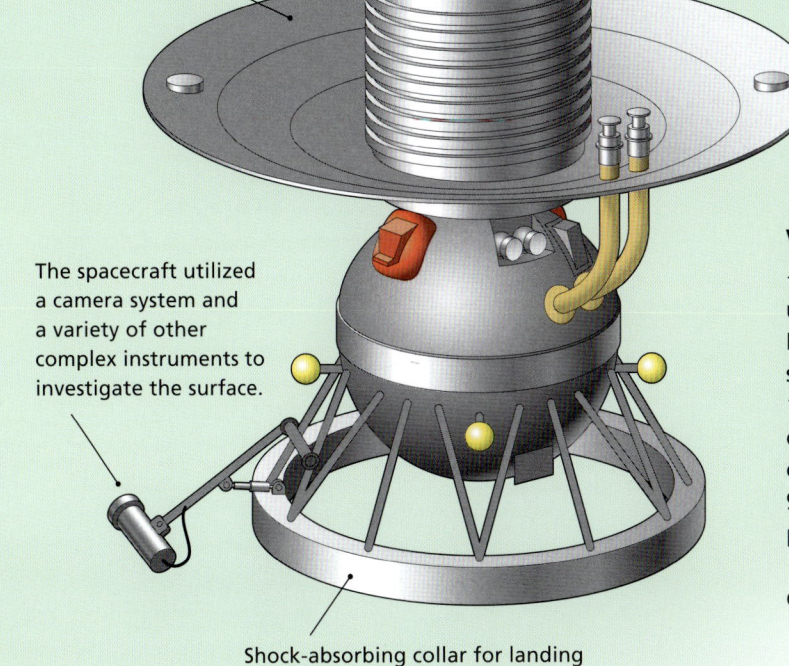

Aerodynamic brake used to assist landing

The spacecraft utilized a camera system and a variety of other complex instruments to investigate the surface.

Shock-absorbing collar for landing

GRAVITATIONAL FIELD The region surrounding a body in which that body's gravitational force can be felt.

LAUNCHER A system of propellant tanks and rocket motors or engines designed to lift a payload into space. It may, or may not, be part of a space vehicle.

MOMENTUM The mass of an object multiplied by its velocity.

PROBE An unmanned spacecraft designed to explore our solar system and beyond.

SLINGSHOT TRAJECTORY A path chosen to use the attractive force of gravity to increase the speed of a spacecraft.

SOLAR PANELS Large flat surfaces covered with thousands of small photoelectric devices that convert solar radiation into electricity.

TRAJECTORY The curved path followed by a projectile.

TRUSS Tubing arrayed in the form of triangles and designed to make a strong frame.

VELOCITY A more precise word to describe how something is moving, because movement has both a magnitude (speed) and a direction.

Venera

◀ The Venera project was a series of unmanned Soviet probes sent to Venus. In February 1966 Venera 3 was the first spacecraft to strike another planet. In 1969 Venera 5 and 6 made soft landings on Venus, but were destroyed by the heat of the planet's surface. In 1975 Venera 9 and 10 transmitted the first closeup photographs of the planet's surface.

This is a diagram of the probe sent down from Venera to the surface.

From Pioneer to Cassini

The journey that caught everyone's imagination was the voyage of two identical probes that were sent to investigate the outer planets. It still fascinates us, for these probes, having completed their original tasks, have continued their journeys and are now beyond the solar system, carrying with them messages from our world in the form of golden disks. They have truly earned the title they were given—Voyager.

Two identical Voyager spacecraft, each with a mass of 815 kilograms, were designed for journeys in which there would be no possibility of repair. They had to be entirely reliable and built to last many years of journey. The communications system, for example, had a complete backup just in case one failed.

Each spacecraft communicates with Earth using a high-gain radio antenna 3.7 meters in diameter that is always facing toward the Earth. The 23-watt radio transmitters are able to send signals over a distance of 1 billion kilometers.

Pioneer

▼ The Pioneer probes were mainly sent to other parts of the solar system. In 1965 Pioneer 6 orbited the Sun to find out what space was like close to the Sun. From it important information was gathered about **SOLAR WINDS** and cosmic rays.

In 1972 Pioneer 10 was the first spacecraft to travel through the **ASTEROID BELT**. Later it flew by Jupiter, discovering its vast magnetism. In 1973 Pioneer 11 went close to Jupiter and then on to Saturn in September 1979. Its photographs showed two additional rings around the planet.

In 1978 Pioneers Venus 1 and 2 orbited Venus, the second probe making **RADAR** images that allowed scientists to see the whole planet's landscape for the first time.

The Pioneers were the first of the space probes that continued into deep space. Pioneer's last, very weak signal was received on January 23, 2003, only because its power source had given out and was no longer sending enough power to the transmitter. Pioneer 10 is now over 12 billion km away. Pioneer 10 will continue to travel in the direction of the red star Aldebaran, which forms the eye of Taurus (the Bull). Aldebaran is about 68 light-years away, and it will take Pioneer over 2 million years to reach it.

The Pioneer 11 mission ended on September 30, 1995, when the last transmission from the space probe was received. The Earth's motion had carried it out of the range of the spacecraft antenna. The space probe is headed toward the constellation of Aquila (the Eagle). Pioneer 11 will pass near one of the stars in the constellation in about four million years.

Communications antennae

Scientific instruments

Power source (radioisotope thermoelectric generator) converts heat from decay of plutonium source into electricity.

54

OUTER WORLDS

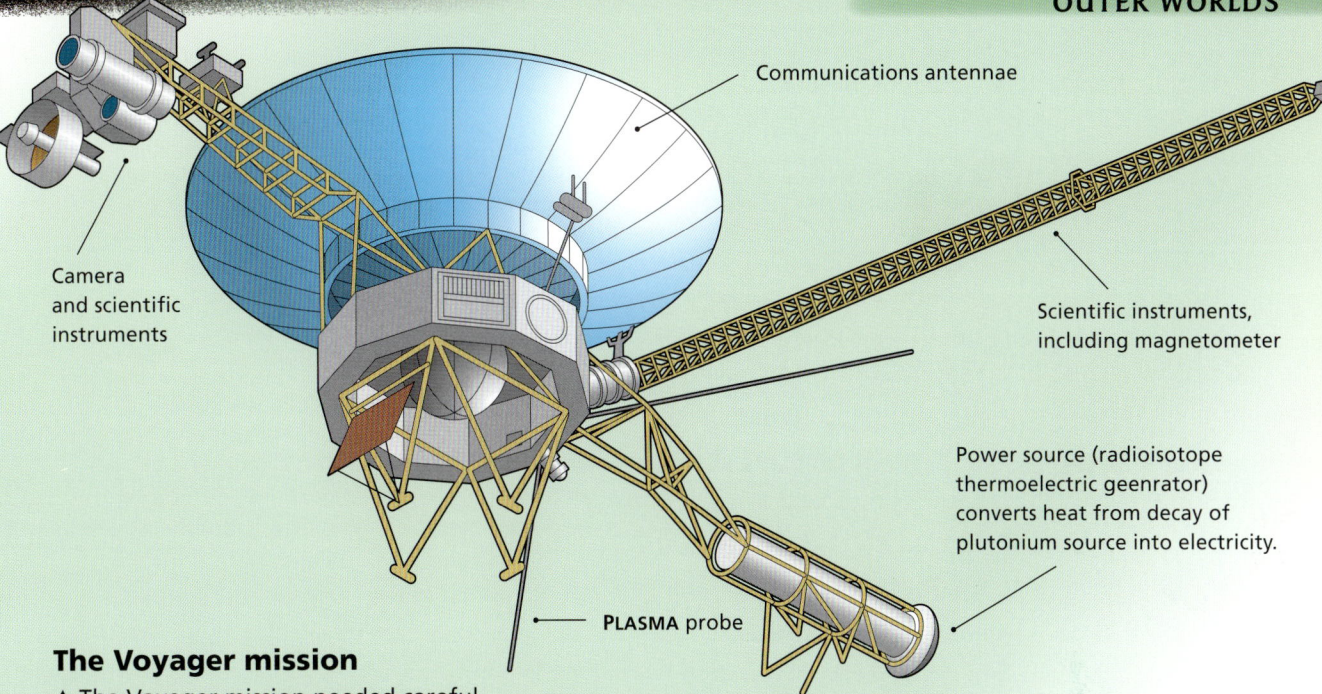

Communications antennae

Camera and scientific instruments

Scientific instruments, including magnetometer

Power source (radioisotope thermoelectric geenrator) converts heat from decay of plutonium source into electricity.

PLASMA probe

The Voyager mission
▲ The Voyager mission needed careful planning. Much had to be learned about how space probes were affected by **RADIATION** and **MICROMETEORITES**, for example. So, before Voyager probes were sent, Pioneer probes were designed and sent off as test vehicles; but they performed a wide range of important scientific explorations in their own right (page 54).

ASTEROID BELT The collection of asteroids that orbit the Sun between the orbits of Mars and Jupiter.

MAGNETIC FIELD The region of influence of a magnetic body.

MICROMETEORITES Tiny pieces of space dust moving at high speeds.

PLASMA A collection of charged particles that behaves something like a gas. It can conduct an electric charge and be affected by magnetic fields.

RADAR Short for radio detecting and ranging. A system of bouncing radio waves from objects in order to map their surfaces and find out how far away they are.

RADIATION The transfer of energy in the form of waves (such as light and heat) or particles (such as from radioactive decay of a material).

SOLAR WIND The flow of tiny charged particles (called plasma) outward from the Sun.

The power for each Voyager spacecraft is supplied by three generators that produce about 400 watts of electrical power. Rather than being instructed directly by ground controllers (as were the Pioneers), the Voyager control systems accept precoded sets of several thousand instructions that can provide automatic operation for days or weeks at a time. These systems also include elaborate error detection and correction routines so that the spacecraft can locate and fix problems before ground controllers are aware of them.

This pair of space probes explored four planets—Jupiter, Saturn, Uranus, and Neptune—as well as dozens of their moons, rings, and **MAGNETIC FIELDS**.

The Voyagers were designed to take advantage of a rare geometric arrangement of the outer planets that occurs only once every 176 years. This placement allows a single space probe to be slingshot by all four gas giants without the need for large on-board propulsion systems. Without these "gravity assists" the flight time to Neptune would have been 30 years.

These space probes took far better pictures of the gas planets, and especially Jupiter and its moons, than had previously been possible.

The success of the Voyager missions encouraged further explorations, and so the Galileo mission was developed. It was no longer possible to use Jupiter as a slingshot, so a three-times orbit around the Sun was used instead, using first Venus and the Earth as gravity assist planets to increase the orbit and speed until the probe could finally be sent to its destination of Jupiter. It was Galileo that orbited Jupiter for 2 years and dropped a probe into the Jovian atmosphere. Cassini, the most recent of the solar system explorers, also used the same gravity-assist trajectory for its launch toward Saturn.

Voyager

Voyager 1, launched on September 5, 1977, flew by Jupiter in March 1979 and reached Saturn in November 1980. Voyager 2, launched on August 20, 1977, moved more slowly, reaching Uranus on January 24, 1986, and Neptune on August 24, 1989.

The launch vehicle for Voyager 1 was a Titan/Centaur. The first stage Titan was powered by both solid and liquid fuel engines. The Centaur stage, 20 meters long and 3 meters in diameter, burned a fuel combination of liquid hydrogen and liquid oxygen. The Titan boosted the Voyager-Centaur combination into low Earth orbit, and the Centaur plus a small solid fuel rocket provided the energy for Voyager 1 to escape Earth orbit.

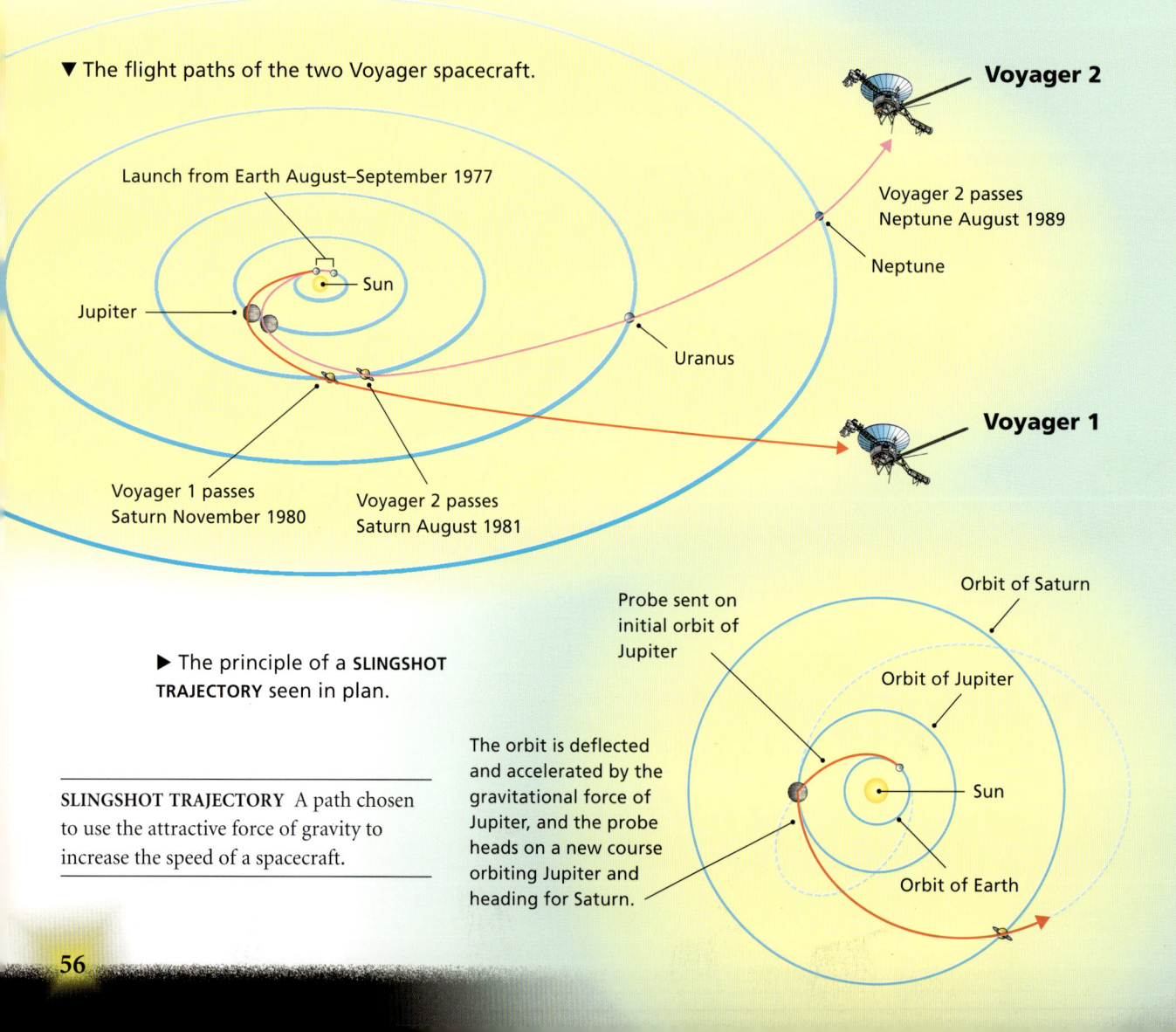

▼ The flight paths of the two Voyager spacecraft.

▶ The principle of a **SLINGSHOT TRAJECTORY** seen in plan.

SLINGSHOT TRAJECTORY A path chosen to use the attractive force of gravity to increase the speed of a spacecraft.

The orbit is deflected and accelerated by the gravitational force of Jupiter, and the probe heads on a new course orbiting Jupiter and heading for Saturn.

OUTER WORLDS

Galileo

The space probe Galileo was sent to explore the planet Jupiter. It arrived in 1995. Its 11 instruments were designed to get more detailed information than could be provided by the earlier Pioneer and Voyager spacecraft. Galileo also carried a probe designed to enter Jupiter's atmosphere, but it was destroyed as it descended.

▼ The flight paths of the Galileo spacecraft using the Sun as a slingshot.

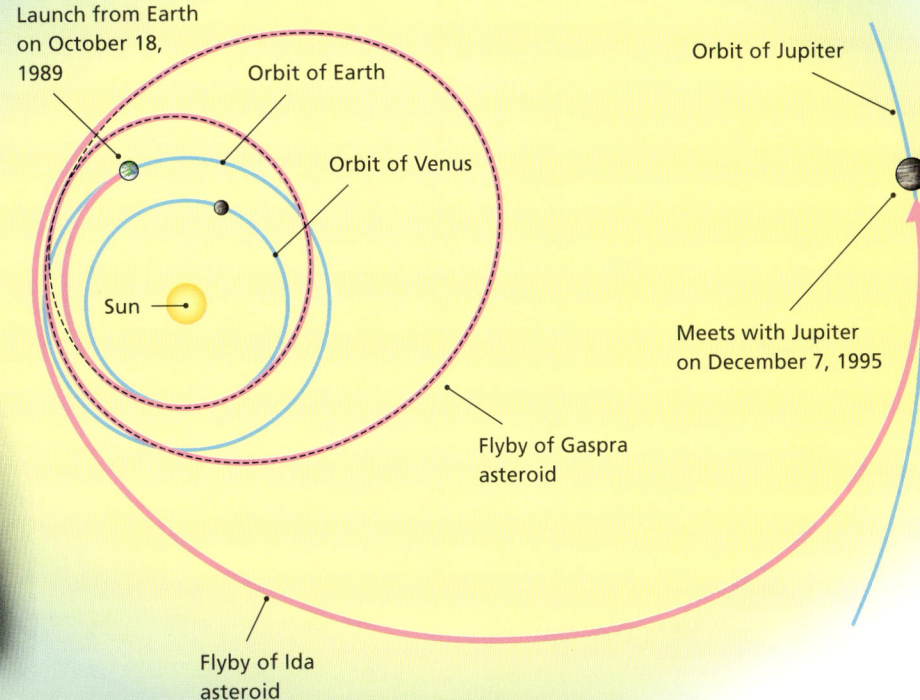

- Launch from Earth on October 18, 1989
- Orbit of Earth
- Orbit of Venus
- Sun
- Orbit of Jupiter
- Meets with Jupiter on December 7, 1995
- Flyby of Gaspra asteroid
- Flyby of Ida asteroid

Cassini-Huygens

This probe was launched in 1997 (see page 21) and designed to reach Saturn in July 2004. The Cassini orbiter is programmed to orbit Saturn and its moons for 4 years, and the Huygens probe will descend into the atmosphere of Titan and land on its surface.

▼ An artist's rendering of the Cassini-Huygens probe journeying through space.

▲ The Cassini-Huygens probe under test.

57

SET GLOSSARY

Metric (SI) and U.S. standard unit conversion table

1 kilometer = 0.6 mile
1 meter = 3.3 feet or 1.1 yards
1 centimeter = 0.4 inch
1 tonne = (metric ton) 1,000 kilograms, or 2,204.6 pounds
1 kilogram = 2.2 pounds
1 gram = 0.035 ounces

1 mile = 1.6 kilometers
1 foot = 0.3 meter
1 inch = 25.4 millimeters or 2.54 centimeters
1 ton = 907.18 kilograms in standard units, 1,016.05 kilograms in metric
1 pound = 0.45 kilograms
1 ounce = 28 grams

0°C = 5/9 (32°F)

ABSOLUTE ZERO The coldest possible temperature, defined as 0 K or −273°C. *See also:* **K**.

ACCELERATE To gain speed.

AERODYNAMIC A shape offering as little resistance to the air as possible.

AIR RESISTANCE The frictional drag that an object creates as it moves rapidly through the air.

AMINO ACIDS Simple organic molecules that can be building blocks for living things.

ANNULAR Ringlike.
An annular eclipse occurs when the dark disk of the Moon does not completely obscure the Sun.

ANTENNA (pl. **ANTENNAE**) A device, often in the shape of a rod or wire, used for sending out and receiving radio waves.

ANTICLINE An arching fold of rock layers where the rocks slope down from the crest.

ANTICYCLONE A roughly circular region of the atmosphere that is spiraling outward and downward.

APOGEE The point on an orbit where the orbiting object is at its farthest from the object it is orbiting.

APOLLO The program developed in the United States by NASA to get people to the Moon's surface and back safely.

ARRAY A regular group or arrangement.

ASH Fragments of lava that have cooled and solidified between when they leave a volcano and when they fall to the surface.

ASTEROID Any of the many small objects within the solar system.
Asteroids are rocky or metallic and are conventionally described as significant bodies with a diameter smaller than 1,000 km. Asteroids mainly occupy a belt between Mars and Jupiter (asteroid belt).

ASTEROID BELT The collection of asteroids that orbit the Sun between the orbits of Mars and Jupiter.

ASTHENOSPHERE The region below the lithosphere, and therefore part of the upper mantle, in which some material may be molten.

ASTRONOMICAL UNIT (**AU**) The average distance from the Earth to the Sun (149,597,870 km).

ASTRONOMY The study of space beyond the Earth and its contents. It includes those phenomena that affect the Earth but that originate in space, such as meteorites and aurora.

ASTROPHYSICS The study of physics in space, what other stars, galaxies, and planets are like, and the physical laws that govern them.

ASYNCHRONOUS Not connected in time or pace.

ATMOSPHERE The envelope of gases that surrounds the Earth and other bodies in the universe.
The Earth's atmosphere is very different from that of other planets, being, for example, far lower in hydrogen and helium than the gas giants and lower in carbon dioxide than Venus, but richer in oxygen than all the others.

ATMOSPHERIC PRESSURE The pressure on the gases in the atmosphere caused by gravity pulling them toward the center of a celestial body.

ATOM The smallest particle of an element.

ATOMIC MASS UNIT A measure of the mass of an atom or molecule.
An atomic mass unit equals one-twelfth of the mass of an atom of carbon-12.

ATOMIC WEAPONS Weapons that rely on the violent explosive force achieved when radioactive materials are made to go into an uncontrollable chain reaction.

ATOMIC WEIGHT The ratio of the average mass of a chemical element's atoms to carbon-12.

AURORA A region of illumination, often in the form of a wavy curtain, high in the atmosphere of a planet.
It is the result of the interaction of the planet's magnetic field with the particles in the solar wind. High-energy electrons from the solar wind race along the planet's magnetic field into the upper atmosphere. The electrons excite atmospheric gases, making them glow.

AXIS (pl. **AXES**) The line around which a body spins.
The Earth spins around an axis through its north and south geographic poles.

BALLISTIC MISSILE A rocket that is guided up in a high arching path; then the fuel supply is cut, and it is allowed to fall to the ground.

BASIN A large depression in the ground (bigger than a crater).

BIG BANG The theory that the universe as we know it started from a single point (called a singularity) and then exploded outward. It is still expanding today.

BINARY STAR A pair of stars that are gravitationally attracted, and that revolve around one another.

BLACK DWARF A degenerate star that has cooled so that it is now not visible.

BLACK HOLE An object that has a gravitational pull so strong that nothing can escape from it.
A black hole may have a mass equal to thousands of stars or more.

BLUE GIANT A young, extremely bright and hot star of very large mass that has used up all its hydrogen and is no longer in the main sequence. When a blue giant ages, it becomes a red giant.

BOILING POINT The change of state of a substance in which a liquid rapidly turns into a gas without a change in temperature.

BOOSTER POD A form of housing that stands outside the main body of the launcher.

CALDERA A large pit in the top of a volcano produced when the top of the volcano explodes and collapses in on itself.

CAPSULE A small pressurized space vehicle.

CATALYST A substance that speeds up a chemical reaction but that is itself unchanged.

CELESTIAL Relating to the sky above, the "heavens."

CENTER OF GRAVITY The point at which all of the mass of an object can be balanced.

CENTRIFUGAL FORCE A force that acts on an orbiting or spinning body, tending to oppose gravity and move away from the center of rotation.
For orbiting objects the centrifugal force acts in the opposite direction from gravity. When satellites orbit the Earth, the centrifugal force balances out the force of gravity.

CENTRIFUGE An instrument for spinning small samples very rapidly.

CHAIN REACTION A sequence of related events with one event triggering the next.

CHASM A deep, narrow trench.

CHROMOSPHERE The shell of gases that makes up part of the atmosphere of a star and lies between the photosphere and the corona.

CIRCUMFERENCE The distance around the edge of a circle or sphere.

COMA The blurred image caused by light bouncing from a collection of dust and ice particles escaping from the nucleus of a comet.

The coma changes the appearance of a comet from a point source of reflective light to a blurry object with a tail.

COMBUSTION CHAMBER A vessel inside an engine or motor where the fuel components mix and are set on fire, that is, they are burned (combusted).

COMET A small object, often described as being like a dirty snowball, that appears to be very bright in the night sky and has a long tail when it approaches the Sun.

Comets are thought to be some of the oldest objects in the solar system.

COMPLEMENTARY COLOR A color that is diametrically opposed in the range, or circle, of colors in the spectrum; for example, cyan (blue) is the complement of red.

COMPOSITE A material made from solid threads in a liquid matrix that is allowed to set.

COMPOUND A substance made from two or more elements that have chemically combined.

Ammonia is an example of a compound made from the elements hydrogen and nitrogen.

CONDENSE/CONDENSATION (1) To make something more concentrated or compact.

(2) The change of state from a gas or vapor to a liquid.

CONDUCTION The transfer of heat between two objects when they touch.

CONSTELLATION One of many commonly recognized patterns of stars in the sky.

CONVECTION/CONVECTION CURRENTS The circulating flow in a fluid (liquid or gas) that occurs when it is heated from below.

Convective flow is caused in a fluid by the tendency for hotter, and therefore less dense, material to rise and for colder, and therefore more dense, material to sink with gravity. That results in a heat transfer.

CORE The central region of a body.

The core of the Earth is about 3,300 km in radius, compared with the radius of the whole Earth, which is 6,300 km.

CORONA (pl. **CORONAE**) (1) A colored circle seen around a bright object such as a star.

(2) The gases surrounding a star such as the Sun. In the case of the Sun and certain other stars these gases are extremely hot.

(3) A circular to oval pattern of faults, fractures, and ridges with a sagging center as found on Venus. In the case of Venus they are a few hundred kilometers in diameter.

CORONAL MASS EJECTIONS Very large bubbles of plasma escaping into the corona.

CORROSIVE SUBSTANCE Something that chemically eats away something else.

COSMOLOGICAL PRINCIPLE States that the way you see the universe is independent of the place where you are (your location). In effect, it means that the universe is roughly uniform throughout.

COSMONAUT A Russian space person.

COSMOS The universe and everything in it. The word "cosmos" suggests that the universe operates according to orderly principles.

CRATER A deep bowl-shaped depression in the surface of a body formed by the high-speed impact of another, smaller body.

Most craters are formed by the impact of asteroids and meteoroids. Craters have both a depression, or pit, and also an elevated rim formed of the material displaced from the central pit.

CRESCENT The appearance of the Moon when it is between a new Moon and a half Moon.

CRUST The solid outer surface of a rocky body.

The crust of the Earth is mainly just a few tens of kilometers thick, compared to the total radius of 6,300 km for the whole Earth. It forms much of the lithosphere.

CRYSTAL An ordered arrangement of molecules in a compound. Crystals that grow freely develop flat surfaces.

CYCLONE A large storm in which the atmosphere spirals inward and upward.

On Earth cyclones have a very low atmospheric pressure at their center and often contain deep clouds.

DARK MATTER Matter that does not shine or reflect light.

No one has ever found dark matter, but it is thought to exist because the amount of ordinary matter in the universe is not enough to account for many gravitational effects that have been observed.

DENSITY A measure of the amount of matter in a space.

Density is often measured in grams per cubic centimeter. The density of the Earth is 5.5 grams per cubic centimeter.

DEORBIT To move out of an orbital position and begin a reentry path toward the Earth.

DEPRESSION (1) A sunken area or hollow in a surface or landscape.

(2) A region of inward swirling air in the atmosphere associated with cloudy weather and rain.

DIFFRACTION The bending of light as it goes through materials of different density.

DISK A shape or surface that looks round and flat.

DOCK To meet with and attach to another space vehicle.

DOCKING PORT/STATION A place on the side of a spacecraft that contains some form of anchoring mechanism and an airlock.

DOPPLER EFFECT The apparent change in pitch of a fast-moving object as it approaches or leaves an observer.

DOWNLINK A communication to Earth from a spacecraft.

DRAG A force that hinders the movement of something.

DWARF STAR A star that shines with a brightness that is average or below.

EARTH The third planet from the Sun and the one on which we live.

The Earth belongs to the group of rocky planets. It is unique in having an oxygen-rich atmosphere and water, commonly found in its three phases—solid, liquid, and gas.

EARTHQUAKE The shock waves produced by the sudden movement of two pieces of brittle crust.

ECCENTRIC A noncircular, or oval, orbit.

ECLIPSE The time when light is cut off by a body coming between the observer and the source of the illumination (for example, eclipse of the Sun), or when the body the observer is on comes between the source of illumination and another body (for example, eclipse of the Moon).

It happens when three bodies are in a line. This phenomenon is not necessarily called an eclipse. Occultations of stars by the Moon and transits of Venus or Mercury are examples of different expressions used instead of "eclipse."

See also: **TOTAL ECLIPSE**.

ECOLOGY The study of living things in their environment.

ELECTRONS Negatively charged particles that are parts of atoms.

ELEMENT A substance that cannot be decomposed into simpler substances by chemical means.

Elements are the building blocks of compounds. For example, silicon and oxygen are elements. They combine to form the compound silicon dioxide, or quartz.

ELLIPTICAL GALAXY A galaxy that has an oval shape rather like a football, and that has no spiral arms.

EL NIÑO A time when ocean currents in the Pacific Ocean reverse from their normal pattern and disrupt global weather patterns. It occurs once every 4 or 5 years.

EMISSION Something that is sent or let out.

ENCKE GAP A gap between rings around Saturn named for the astronomer Johann Franz Encke (1791–1865).

EPOXY RESIN Adhesives that develop their strength as they react, or "cure," after mixing.

EQUATOR The ring drawn around a body midway between the poles.

EQUILIBRIUM A state of balance.

ESA The European Space Agency. ESA is an organizaton of European countries for cooperation in space research and technology. It operates several installations around Europe and has its headquarters in Paris, France.

ESCARPMENT A sharp-edged ridge.

59

EVAPORATE/EVAPORATION The change in state from liquid to a gas.

EXOSPHERE The outer part of the atmosphere starting about 500 km from the surface. This layer contains so little air that molecules rarely collide.

EXTRAVEHICULAR ACTIVITY Any task performed by people outside the protected environment of a space vehicle's pressurized compartments. Extravehicular activities (EVA) include repairing equipment in the Space Shuttle bay.

FALSE COLOR The colors used to make the appearance of some property more obvious.
   They are part of the computer generation of an image.

FAULT A place in the crust where rocks have fractured, and then one side has moved relative to the other.
   A fault is caused by excessive pressure on brittle rocks.

FLUORESCENT Emitting the visible light produced by a substance when it is struck by invisible waves, such as ultraviolet waves.

FRACTURE A break in brittle rock.

FREQUENCY The number of complete cycles of (for example, radio) waves received per second.

FRICTION The force that resists two bodies that are in contact.
   For example, the effect of the ocean waters moving as tides slows the Earth's rotation.

FUSION The joining of atomic nuclei to form heavier nuclei.
   This process results in the release of huge amounts of energy.

GALAXY A system of stars and interstellar matter within the universe.
   Galaxies may contain billions of stars.

GALILEAN SATELLITES The four large satellites of Jupiter discovered by astronomer Galileo Galilei in 1610. They are Callisto, Europa, Ganymede, and Io.

GALILEO A U.S. space probe launched in October 1989 and designed for intensive investigation of Jupiter.

GEIGER TUBE A device to detect radioactive materials.

GEOSTATIONARY ORBIT A circular orbit 35,786 km directly above the Earth's equator.
   Communications satellites frequently use this orbit. A satellite in a geostationary orbit will move at the same rate as the Earth's rotation, completing one revolution in 24 hours. That way it remains at the same point over the Earth's equator.

GEOSTATIONARY SATELLITE A man-made satellite in a fixed or geosynchronous orbit around the Earth.

GEOSYNCHRONOUS ORBIT An orbit in which a satellite makes one circuit of the Earth in 24 hours.
   A geosynchronous orbit coincides with the Earth's orbit—it takes the same time to complete an orbit as it does for the Earth to make one complete rotation. If the orbit is circular and above the equator, then the satellite remains over one particular point of the equator; that is called a geostationary orbit.

GEOSYNCLINE A large downward sag or trench that forms in the Earth's crust as a result of colliding tectonic plates.

GEYSER A periodic fountain of material. On Earth geysers are of water and steam, but on other planets and moons they are formed from other substances, for example, nitrogen gas on Triton.

GIBBOUS When between half and a full disk of a body can be seen lighted by the Sun.

GIMBALS A framework that allows anything inside it to move in a variety of directions.

GLOBAL POSITIONING SYSTEM A network of geostationary satellites that can be used to locate the position of any object on the Earth's surface.

GRANULATION The speckled pattern we see in the Sun's photosphere as a result of convectional overturning of gases.

GRAVITATIONAL FIELD The region surrounding a body in which that body's gravitational force can be felt.
   The gravitational field of the Sun spreads over the entire solar system. The gravitational fields of the planets each exert some influence on the orbits of their neighbors.

GRAVITY/GRAVITATIONAL FORCE/ GRAVITATIONAL PULL The force of attraction between bodies. The larger an object, the more its gravitational pull on other objects.
   The Sun's gravity is the most powerful in the solar system, keeping all of the planets and other materials within the solar system.

GREAT RED SPOT A large, almost permanent feature of the Jovian atmosphere that moves around the planet at about latitude 23°S.

GREENHOUSE EFFECT The increase in atmospheric temperature produced by the presence of carbon dioxide in the air.
   Carbon dioxide has the ability to soak up heat radiated from the surface of a planet and partly prevent its escape. The effect is similar to that produced by a greenhouse.

GROUND STATION A receiving and transmitting station in direct communication with satellites. Such stations are characterized by having large dish-shaped antennae.

GULLY (pl. **GULLIES**) A trench in the land surface formed, on Earth, by running water.

GYROSCOPE A device in which a rapidly spinning wheel is held in a frame in such a way that it can rotate in any direction. The momentum of the wheel means that the gyroscope retains its position even when the frame is tilted.

HEAT SHIELD A protective device on the outside of a space vehicle that absorbs the heat during reentry and protects it from burning up.

HELIOPAUSE The edge of the heliosphere.

HELIOSEISMOLOGY The study of the internal structure of the Sun by modeling the Sun's patterns of internal shock waves.

HELIOSPHERE The entire range of influence of the Sun. It extends to the edge of the solar system.

HUBBLE SPACE TELESCOPE An orbiting telescope (and so a satellite) that was placed above the Earth's atmosphere so that it could take images that were far clearer than anything that could be obtained from the surface of the Earth.

HURRICANE A very violent cyclone that begins close to the equator, and that contains winds of over 117 km/hr.

ICE CAP A small mountainous region that is covered in ice.

INFRARED Radiation with a wavelength that is longer than red light.

INNER PLANETS The rocky planets closest to the Sun. They are Mercury, Venus, Earth, and Mars.

INTERNATIONAL SPACE STATION The international orbiting space laboratory.

INTERPLANETARY DUST The fine dustlike material that lies scattered through space, and that exists between the planets as well as in outer space.

INTERSTELLAR Between the stars.

IONIZED Matter that has been converted into small charged particles called ions.
   An atom that has gained or lost an electron.

IONOSPHERE A part of the Earth's atmosphere in which the number of ions (electrically charged particles) is enough to affect how radio waves move.
   The ionosphere begins about 50 km above the Earth's surface.

IRREGULAR SATELLITES Satellites that orbit in the opposite direction from their parent planet.
   This motion is also called retrograde rotation.

ISOTOPE Atoms that have the same number of protons in their nucleus, but that have different masses; for example, carbon-12 and carbon-14.

JOVIAN PLANETS An alternative group name for the gas giant planets: Jupiter, Saturn, Uranus, and Neptune.

JUPITER The fifth planet from the Sun and two planets farther away from the Sun than the Earth.
   Jupiter is 318 times as massive as the Earth and 1,500 times as big by volume. It is the largest of the gas giants.

K Named for British scientist Lord Kelvin (1824–1907), it is a measurement of absolute temperature. Zero K is called absolute zero and is only approached in deep space: ice melts at 273 K, and water boils at 373 K.

KEELER GAP A gap in the rings of Saturn named for the astronomer James Edward Keeler (1857–1900).

KILOPARSEC A unit of a thousand parsecs. A parsec is the unit used for measuring the largest distances in the universe.

KUIPER BELT A belt of planetesimals (small rocky bodies, one kilometer to hundreds of kilometers across) much closer to the Sun than the Oort cloud.

LANDSLIDE A sudden collapse of material on a steep slope.

LA NIÑA Below normal ocean temperatures in the eastern Pacific Ocean that disrupt global weather patterns.

LATITUDE Angular distance north or south of the equator, measured through 90°.

LAUNCH VEHICLE/LAUNCHER A system of propellant tanks and rocket motors or engines designed to lift a payload into space. It may, or may not, be part of a space vehicle.

LAVA Hot, melted rock from a volcano.
   Lava flows onto the surface of a planet and cools and hardens to form new rock. Most of the lava on Earth is made of basalt.

LAVA FLOW A river or sheet of liquid volcanic rock.

LAWS OF MOTION Formulated by Sir Isaac Newton, they describe the forces that act on a moving object.
   The first law states that an object will keep moving in a straight line at constant speed unless it is acted on by a force.
   The second law states that the force on an object is related to the mass of the object multiplied by its acceleration.
   The third law states that an action always has an equal and directly opposite reaction.

LIFT An upthrust on the wing of a plane that occurs when it moves rapidly through the air. It is the main way of suspending an airplane during flight. The engines simply provide the forward thrust.

LIGHT-YEAR The distance traveled by light through space in one Earth year, or 63,240 astronomical units.
   The speed of light is the speed that light travels through a vacuum, which is 299,792 km/s.

LIMB The outer edge of a celestial body, including an atmosphere if it has one.

LITHOSPHERE The upper part of the Earth, corresponding generally to the crust and believed to be about 80 km thick.

LOCAL GROUP The Milky Way, the Magellanic Clouds, the Andromeda Galaxy, and over 20 other relatively near galaxies.

LUNAR Anything to do with the Moon.

MAGELLANIC CLOUD Either of two small galaxies that are companions to the Milky Way Galaxy.

MAGMA Hot, melted rock inside the Earth that, when cooled, forms igneous rock.
   Magma is associated with volcanic activity.

MAGNETIC FIELD The region of influence of a magnetic body.
   The Earth's magnetic field stretches out beyond the atmosphere into space. There it interacts with the solar wind to produce auroras.

MAGNETISM An invisible force that has the property of attracting iron and similar metals.

MAGNETOPAUSE The outer edge of the magnetosphere.

MAGNETOSPHERE A region in the upper atmosphere, or around a planet, where magnetic phenomena such as auroras are found.

MAGNITUDE A measure of the brightness of a star.
   The apparent magnitude is the brightness of a celestial object as seen from the Earth. The absolute magnitude is the standardized brightness measured as though all objects were the same distance from the Earth. The brighter the object, the lower its magnitude number. For example, a star of magnitude 4 is 2.5 times as bright as one of magnitude 5. A difference of five magnitudes is the same as a difference in brightness of 100 to 1. The brightest stars have negative numbers. The Sun's apparent magnitude is −26.8. Its absolute magnitude is 4.8.

MAIN SEQUENCE The 90% of stars in the universe that represent the mature phase of stars with small or medium mass.

MANTLE The region of a planet between the core and the crust.
   The Earth's mantle is about 2,900 km thick, and its upper surface may be molten in some places.

MARE (pl. **MARIA**) A flat, dark plain created by lava flows. They were once thought to be seas.

MARS The fourth planet from the Sun in our solar system and one planet farther away from the Sun than the Earth.
   Mars is a rocky planet almost half the diameter of Earth that is a distinctive rust-red color.

MASCON A region of higher surface density on the Moon.

MASS The amount of matter in an object.
   The amount of matter, and so the mass, remains the same, but the effect of gravity gives the mass a weight. The weight depends on the gravitational pull. Thus a ball will have the same mass on the Earth and on the Moon, but it will weigh a sixth as much on the Moon because the force of gravity there is only a sixth as strong.

MATTER Anything that exists in physical form.
   Everything we can see is made of matter. The building blocks of matter are atoms.

MERCURY The closest planet to the Sun in our solar system and two planets closer to the Sun than Earth.
   Mercury is a gray-colored rocky planet less than half the diameter of Earth. It has the most extreme temperature range of any planet in our solar system.

MESOSPHERE One of the upper regions of the atmosphere, beginning at the top of the stratosphere and continuing from 50 km upward until the temperature stops declining.

METEOR A streak of light (shooting star) produced by a meteoroid as it enters the Earth's atmosphere.
   The friction with the Earth's atmosphere causes the small body to glow (become incandescent). That is what we see as a streak of light.

METEORITE A meteor that reaches the Earth's surface.

METEOROID A small body moving in the solar system that becomes a meteor if it enters the Earth's atmosphere.
   Meteoroids are typically only a few millimeters across and burn up as they go through the atmosphere, but some have crashed to the Earth, making large craters.

MICROMETEORITES Tiny pieces of space dust moving at high speeds.

MICRON A millionth of a meter.

MICROWAVELENGTH Waves at the shortest end of the radio wavelengths.

MICROWAVE RADIATION The background radiation that is found everywhere in space, and whose existence is used to support the Big Bang theory.

MILKY WAY The spiral galaxy in which our star and solar system are situated.

MINERAL A solid crystalline substance.

MINOR PLANET Another term for an asteroid.

M NUMBER In 1781 Charles Messier began a catalogue of the objects he could see in the night sky. He gave each of them a unique number. The first entry was called M1. There is no significance to the number in terms of brightness, size, closeness, or otherwise.

MODULE A section, or part, of a space vehicle.

MOLECULE A group of two or more atoms held together by chemical bonds.

MOLTEN Liquid, suggesting that it has changed from a solid.

MOMENTUM The mass of an object multiplied by its velocity.

MOON The natural satellite that orbits the Earth.
   Other planets have large satellites, or moons, but none is relatively as large as our Moon, suggesting that it has a unique origin.

MOON The name generally given to any large natural satellite of a planet.

MOUNTAIN RANGE A long, narrow region of very high land that contains several or many mountains.

NASA The National Aeronautics and Space Administration.
   NASA was founded in 1958 for aeronautical and space exploration. It operates several installations around the country and has its headquarters in Washington, D.C.

61

NEAP TIDE A tide showing the smallest difference between high and low tides.

NEBULA (pl. **NEBULAE**) Clouds of gas and dust that exist in the space between stars.

The word means mist or cloud and is also used as an alternative to galaxy. The gas makes up to 5% of the mass of a galaxy. What a nebula looks like depends on the arrangement of gas and dust within it.

NEPTUNE The eighth planet from the Sun in our solar system and five planets farther away from the Sun than the Earth.

Neptune is a gas planet that is almost four times the diameter of Earth. It is blue.

NEUTRINOS An uncharged fundamental particle that is thought to have no mass.

NEUTRONS Particles inside the core of an atom that are neutral (have no charge).

NEUTRON STAR A very dense star that consists only of tightly packed neutrons. It is the result of the collapse of a massive star.

NOBLE GASES The unreactive gases, such as neon, xenon, and krypton.

NOVA (pl. **NOVAE**) (1) A star that suddenly becomes much brighter, then fades away to its original brightness within a few months. *See also:* **SUPERNOVA**.

(2) A radiating pattern of faults and fractures unique to Venus.

NUCLEAR DEVICES Anything that is powered by a source of radioactivity.

NUCLEUS (pl. **NUCLEI**) The centermost part of something, the core.

OORT CLOUD A region on the edge of the solar system that consists of planetesimals and comets that did not get caught up in planet making.

OPTICAL Relating to the use of light.

ORBIT The path followed by one object as it tracks around another.

The orbits of the planets around the Sun and moons around their planets are oval, or elliptical.

ORGANIC MATERIAL Any matter that contains carbon and is alive.

OUTER PLANETS The gas giant planets Jupiter, Saturn, Uranus, and Neptune plus the rocky planet Pluto.

OXIDIZER The substance in a reaction that removes electrons from and thereby oxidizes (burns) another substance.

In the case of oxygen this results in the other substance combining with the oxygen to form an oxide (also called an oxidizing agent).

OZONE A form of oxygen (O_3) with three atoms in each molecule instead of the more usual two (O_2).

OZONE HOLE The observed lack of the gas ozone in the upper atmosphere.

PARSEC The unit used for measuring the largest distances in the universe.

A parsec is the distance at which an observer in space would see the radius of the orbit as making one second of arc. This gives a distance of about 3.26 light-years.
See also: **KILOPARSEC**.

PAYLOAD The spacecraft that is carried into space by a launcher.

PENUMBRA (1) A region that is in semidarkness during an eclipse.

(2) The part of a sunspot surrounding the umbra.

PERCOLATE To flow by gravity between particles, for example, of soil.

PERIGEE The point on an orbit where the orbiting object is as close as it ever comes to the object it is orbiting.

PHARMACEUTICAL Relating to medicinal drugs.

PHASE The differing appearance of a body that is closer to the Sun, and that is illuminated by it.

PHOTOCHEMICAL SMOG A hazy atmosphere, often brown, resulting from the reaction of nitrogen gases with sunlight.

PHOTOMOSAIC A composite picture made up of several other pictures that individually only cover a small area.

PHOTON A particle (quantum) of electromagnetic radiation.

PHOTOSPHERE A shell of the Sun that we regard as its visible surface.

PHOTOSYNTHESIS The process that plants use to combine the substances in the environment, such as carbon dioxide, minerals, and water, with oxygen and energy-rich organic compounds by using the energy of sunlight.

PIONEER A name for a series of unmanned U.S. spacecraft.

Pioneer 1 was launched into lunar orbit on October 11, 1958. The others all went into deep space.

PLAIN A flat or gently rolling part of a landscape.

Plains are confined to lowlands. If a flat surface exists in an upland, it is called a plateau.

PLANE A flat surface.

PLANET Any of the large bodies that orbit the Sun.

The planets are (outward from the Sun): Mercury, Venus, Earth, Mars, Jupiter, Saturn, Uranus, Neptune, and Pluto. The rocky planets all have densities greater than 3 grams per cubic centimeter; the gaseous ones less than 2 grams per cubic centimeter.

PLANETARY NEBULA A compact ring or oval nebula that is made of material thrown out of a hot star.

The term "planetary nebula" is a misnomer; dying stars create these cocoons when they lose outer layers of gas. The process has nothing to do with planet formation, which is predicted to happen early in a star's life.

The term originates from a time when people, looking through weak telescopes, thought that the nebulae resembled planets within the solar system, when in fact they were expanding shells of glowing gas in far-off galaxies.

PLANETESIMAL Small rocky bodies one kilometer to hundreds of kilometers across.

The word especially relates to materials that exist in the early stages of the formation of a star and its planets from the dust of a nebula, which will eventually group together to form planets. Some are rock, others a mixture of rock and ice.

PLANKTON Microscopic creatures that float in water.

PLASMA A collection of charged particles that behaves something like a gas. It can conduct an electric charge and be affected by magnetic fields.

PLASTIC The ability of certain solid substances to be molded or deformed to a new shape under pressure without cracking.

PLATE A very large unbroken part of the crust of a planet. Also called tectonic plate.

On Earth the tectonic plates are dragged across the surface by convection currents in the underlying mantle.

PLATEAU An upland plain or tableland.

PLUTO The ninth planet from the Sun and six planets farther from the Sun than the Earth.

Pluto is one of the rocky planets, but it is very different from the others, perhaps being a mixture of rock and ice. It is about two-thirds the size of our Moon.

POLE The geographic pole is the place where a line drawn along the axis of rotation exits from a body's surface.

Magnetic poles do not always correspond with geographic poles.

POLYMER A compound that is made up of long chains formed by combining molecules called monomers as repeating units. ("Poly" means many, "mer" means part.)

PRESSURE The force per unit area.

PROBE An unmanned spacecraft designed to explore our solar system and beyond.

Voyager, Cassini, and Magellan are examples of probes.

PROJECTILE An object propelled through the air or space by an external force or an on-board engine.

PROMINENCE A cloud of burning ionized gas that rises through the Sun's chromosphere into the corona. It can take the form of a sheet or a loop.

PROPELLANT A gas, liquid, or solid that can be expelled rapidly from the end of an object in order to give it motion.

Liquefied gases and solids are used as rocket propellants.

PROPULSION SYSTEM The motors or rockets and their tanks designed to give a launcher or space vehicle the thrust it needs.

PROTEIN Molecules in living things that are vital for building tissues.

PROTONS Positively charged particles from the core of an atom.

PROTOSTAR A cloud of gas and dust that begins to swirl around; the resulting gravity gives birth to a star.

PULSAR A neutron star that is spinning around, releasing electromagnetic radiation, including radio waves.

QUANTUM THEORY A concept of how energy can be divided into tiny pieces called quanta, which is the key to how the smallest particles work and how they build together to make the universe around us.

QUASAR A rare starlike object of enormous brightness that gives out radio waves, which are thought to be released as material is sucked toward a black hole.

RADAR Short for radio detecting and ranging. A system of bouncing radio waves from objects in order to map their surfaces and find out how far away they are.

Radar is useful in conditions where visible light cannot be used.

RADIATION/RADIATE The transfer of energy in the form of waves (such as light and heat) or particles (such as from radioactive decay of a material).

RADIOACTIVE/RADIOACTIVITY The property of some materials that emit radiation or energetic particles from the nucleus of their atoms.

RADIOACTIVE DECAY The change that takes place inside radioactive materials and causes them to give out progressively less radiation over time.

RADIO GALAXY A galaxy that gives out radio waves of enormous power.

RADIO INTERFERENCE Reduction in the radio communication effectiveness of the ionosphere caused by sunspots and other increases in the solar wind.

RADIO TELESCOPE A telescope that is designed to detect radio waves rather than light waves.

RADIO WAVES A form of electromagnetic radiation, like light and heat. Radio waves have a longer wavelength than light waves.

RADIUS (pl. **RADII**) The distance from the center to the outside of a circle or sphere.

RAY A line across the surface of a planet or moon made by material from a crater being flung across the surface.

REACTION An opposition to a force.

REACTIVE The ability of a chemical substance to combine readily with other substances. Oxygen is an example of a reactive substance.

RED GIANT A cool, large, bright star at least 25 times the diameter of our Sun.

REFLECT/REFLECTION/REFLECTIVE To bounce back any light that falls on a surface.

REGULAR SATELLITES Satellites that orbit in the same direction as their parent planet. This motion is also called synchronous rotation.

RESOLVING POWER The ability of an optical telescope to form an image of a distant object.

RETROGRADE DIRECTION An orbit the opposite of normal—that is, a planet that spins so the Sun rises in the west and sinks in the east.

RETROROCKET A rocket that fires against the direction of travel in order to slow down a space vehicle.

RIDGE A narrow crest of an upland area.

RIFT A trench made by the sinking of a part of the crust between parallel faults.

RIFT VALLEY A long trench in the surface of a planet produced by the collapse of the crust in a narrow zone.

ROCKET Any kind of device that uses the principle of jet propulsion, that is, the rapid release of gases designed to propel an object rapidly.

The word is also applied loosely to fireworks and spacecraft launch vehicles.

ROCKET ENGINE A propulsion system that burns liquid fuel such as liquid hydrogen.

ROCKET MOTOR A propulsion system that burns solid fuel such as hydrazine.

ROCKETRY Experimentation with rockets.

ROTATION Spinning around an axis.

SAND DUNE An aerodynamically shaped hump of sand.

SAROS CYCLE The interval of 18 years $11^{1}/_{3}$ days needed for the Earth, Sun, and Moon to come back into the same relative positions. It controls the pattern of eclipses.

SATELLITE (1) An object that is in an orbit around another object, usually a planet.

The Moon is a satellite of the Earth.
See also: **IRREGULAR SATELLITE, MOON, GALILEAN SATELLITE, REGULAR SATELLITE, SHEPHERD SATELLITE.**

(2) A man-made object that orbits the Earth. Usually used as a term for an unmanned spacecraft whose job is to acquire or transfer data to and from the ground.

SATURN The sixth planet from the Sun and three planets farther away from the Sun than the Earth.

It is the least-dense planet in the solar system, having 95 times the mass of the Earth, but 766 times the volume. It is one of the gas giant planets.

SCARP The steep slope of a sharp-crested ridge.

SEASONS The characteristic cycle of events in the heating of the Earth that causes related changes in weather patterns.

SEDIMENT Any particles of material that settle out, usually in layers, from a moving fluid such as air or water.

SEDIMENTARY Rocks deposited in layers.

SEISMIC Shaking, relating to earthquakes.

SENSOR A device used to detect something. Your eyes, ears, and nose are all sensors. Satellites use sensors that mainly detect changes in radio and other waves, including sunlight.

SHEPHERD SATELLITES Larger natural satellites that have an influence on small debris in nearby rings because of their gravity.

SHIELD VOLCANO A volcanic cone that is broad and gently sloping.

SIDEREAL MONTH The average time that the Moon takes to return to the same position against the background of stars.

SILT Particles with a range of 2 microns to 60 microns across.

SLINGSHOT TRAJECTORY A path chosen to use the attractive force of gravity to increase the speed of a spacecraft.

The craft is flown toward the planet or star, and it speeds up under the gravitational force. At the correct moment the path is taken to send the spacecraft into orbit and, when pointing in the right direction, to turn it from orbit, with its increased velocity, toward the final destination.

SOLAR Anything to do with the Sun.

SOLAR CELL A photoelectric device that converts the energy from the Sun (solar radiation) into electrical energy.

SOLAR FLARE Any sudden explosion from the surface of the Sun that sends ultraviolet radiation into the chromosphere. It also sends out some particles that reach Earth and disrupt radio communications.

SOLAR PANELS Large flat surfaces covered with thousands of small photoelectric devices that convert solar radiation into electricity.

SOLAR RADIATION The light and heat energy sent into space from the Sun.

Visible light and heat are just two of the many forms of energy sent by the Sun to the Earth.

SOLAR SYSTEM The Sun and the bodies orbiting around it.

The solar system contains nine major planets, at least 60 moons (large natural satellites), and a vast number of asteroids and comets, together with the gases within the system.

SOLAR WIND The flow of tiny charged particles (called plasma) outward from the Sun.

The solar wind stretches out across the solar system.

SONIC BOOM The noise created when an object moves faster than the speed of sound.

SPACE Everything beyond the Earth's atmosphere.

The word "space" is used rather generally. It can be divided up into inner space—the solar system, and outer space—everything beyond the solar system, for example, interstellar space.

SPACECRAFT Anything capable of moving beyond the Earth's atmosphere. Spacecraft can be manned or unmanned. Unmanned spacecraft are often referred to as space probes if they are exploring new areas.

SPACE RACE The period from the 1950s to the 1970s when the United States and the Soviet Union competed to be first in achievements in space.

63

SPACE SHUTTLE NASA's reusable space vehicle that is launched like a rocket but returns like a glider.

SPACE STATION A large man-made satellite used as a base for operations in space.

SPEED OF LIGHT See: **LIGHT-YEAR**.

SPHERE A ball-shaped object.

SPICULES Jets of relatively cool gas that move upward through the chromosphere into the corona.

SPIRAL GALAXY A galaxy that has a core of stars at the center of long curved arms made of even more stars arranged in a spiral shape.

SPRING TIDE A tide showing the greatest difference between high and low tides.

STAR A large ball of gases that radiates light. The star nearest the Earth is the Sun.

There are enormous numbers of stars in the universe, but few can be seen with the naked eye. Stars may occur singly, as our Sun, or in groups, of which pairs are most common.

STAR CLUSTER A group of gravitationally connected stars.

STELLAR WIND The flow of tiny charged particles (called plasma) outward from a star.

In our solar system the stellar wind is the same as the solar wind.

STRATOSPHERE The region immediately above the troposphere where the temperature increases with height, and the air is always stable.

It acts like an invisible lid, keeping the clouds in the troposphere.

SUBDUCTION ZONES Long, relatively thin, but very deep regions of the crust where one plate moves down and under, or subducts, another. They are the source of mountain ranges.

SUN The star that the planets of the solar system revolve around.

The Sun is 150 million km from the Earth and provides energy (in the form of light and heat) to our planet. Its density of 1.4 grams per cubic centimeter is similar to that of a gas giant planet.

SUNSPOT A spiral of gas found on the Sun that is moving slowly upward, and that is cooler than the surrounding gas and so looks darker.

SUPERNOVA A violently exploding star that becomes millions or even billions of times brighter than when it was younger and stable.

See also: **NOVA**.

SYNCHRONOUS Taking place at the same time.

SYNCHRONOUS ORBIT An orbit in which a satellite (such as a moon) moves around a planet in the same time that it takes for the planet to make one rotation on its axis.

SYNCHRONOUS ROTATION When two bodies make a complete rotation on their axes in the same time.

As a result, each body always has the same side facing the other. The Moon and Venus are in synchronous rotation with the Earth.

SYNODIC MONTH The complete cycle of phases of the Moon as seen from Earth. It is 29.531 solar days (29 days, 12 hours, 44 minutes, 3 seconds).

SYNODIC PERIOD The time needed for an object within the solar system, such as a planet, to return to the same place relative to the Sun as seen from the Earth.

TANGENT A direction at right angles to a line radiating from a circle or sphere.

If you make a wheel spin, for example, by repeatedly giving it a glancing blow with your hand, the glancing blow is moving along a tangent.

TELECOMMUNICATIONS Sending messages by means of telemetry, using signals made into waves such as radio waves.

THEORY OF RELATIVITY A theory based on how physical laws change when an observer is moving. Its most famous equation says that at the speed of light, energy is related to mass and the speed of light.

THERMOSPHERE A region of the upper atmosphere above the mesosphere.

It absorbs ultraviolet radiation and is where the ionosphere has most effect.

THRUST A very strong and continued pressure.

THRUSTER A term for a small rocket engine.

TIDE Any kind of regular, or cyclic, change that occurs due to the effect of the gravity of one body on another.

We are used to the ocean waters of the Earth being affected by the gravitational pull of the Moon, but tides also cause a small alteration of the shape of a body. This is important in determining the shape of many moons and may even be a source of heating in some.

See also: **NEAP TIDE** and **SPRING TIDE**.

TOPOGRAPHY The shape of the land surface in terms of height.

TOTAL ECLIPSE When one body (such as the Moon or Earth) completely obscures the light source from another body (such as the Earth or Moon).

A total eclipse of the Sun occurs when it is completely blocked out by the Moon.

A total eclipse of the Moon occurs when it passes into the Earth's shadow to such a degree that light from the Sun is completely blocked out.

TRAJECTORY The curved path followed by a projectile.

See also: **SLINGSHOT TRAJECTORY**.

TRANSPONDER Wireless receiver and transmitter.

TROPOSPHERE The lowest region of the atmosphere, where all of the Earth's clouds form.

TRUSS Tubing arrayed in the form of triangles and designed to make a strong frame.

ULTRAVIOLET A form of radiation that is just beyond the violet end of the visible spectrum and so is called "ultra" (more than) violet. At the other end of the visible spectrum is "infra" (less than) red.

UMBRA (1) A region that is in complete darkness during an eclipse.

(2) The darkest region in the center of a sunspot.

UNIVERSE The entirety of everything there is; the cosmos.

Many space scientists prefer to use the term "cosmos," referring to the entirety of energy and matter.

UNSTABLE In atmospheric terms the potential churning of the air in the atmosphere as a result of air being heated from below. There is a chance of the warmed, less-dense air rising through the overlying colder, more-dense air.

UPLINK A communication from Earth to a spacecraft.

URANUS The seventh planet from the Sun and four planets farther from the Sun than the Earth.

Its diameter is four times that of the Earth. It is one of the gas giant planets.

VACUUM A space that is entirely empty.

A vacuum lacks any matter.

VALLEY A natural long depression in the landscape.

VELOCITY A more precise word to describe how something is moving, because movement has both a magnitude (speed) and a direction.

VENT The tube or fissure that allows volcanic materials to reach the surface of a planet.

VENUS The second planet from the Sun and our closest neighbor.

It appears as an evening and morning "star" in the sky. Venus is very similar to the Earth in size and mass.

VOLCANO A mound or mountain that is formed from ash or lava.

VOYAGER A pair of U.S. space probes designed to provide detailed information about the outer regions of the solar system.

Voyager 1 was launched on September 5, 1977. Voyager 2 was launched on August 20, 1977, but traveled more slowly than Voyager 1. Both Voyagers are expected to remain operational until 2020, by which time they will be well outside the solar system.

WATER CYCLE The continuous cycling of water, as vapor, liquid, and solid, between the oceans, the atmosphere, and the land.

WATER VAPOR The gaseous form of water. Also sometimes referred to as moisture.

WEATHERING The breaking down of a rock, perhaps by water, ice, or repeated heating and cooling.

WHITE DWARF Any star originally of low mass that has reached the end of its life.

X-RAY An invisible form of radiation that has extremely short wavelengths just beyond the ultraviolet.

X-rays can go through many materials that light will not.

SET INDEX

Using the set index

This index covers all eight volumes in the *Space Science* set:

Vol. no.	Title
1:	*How the universe works*
2:	*Sun and solar system*
3:	*Earth and Moon*
4:	*Rocky planets*
5:	*Gas giants*
6:	*Journey into space*
7:	*Shuttle to Space Station*
8:	*What satellites see*

An example entry:
Index entries are listed alphabetically.

Moon rover **3**: 48–49, **6**: 51

Volume numbers are in bold and are followed by page references.
   In the example above, "Moon rover" appears in Volume 3: *Earth and Moon* on pages 48–49 and in Volume 6: *Journey into space* on page 51. Many terms are also covered separately in the Glossary on pages 58–64.
   See, see also, or *see under* refers to another entry where there will be additional relevant information.

A

absolute zero **4**: 46, **5**: 54, **7**: 57
accretion (buildup) disk **1**: 36, 37, 39
active galactic nucleus **1**: 36, 37.
   See also black holes
Adrastea (Jupiter moon) **5**: 34, 35
aerodynamic design, rockets **6**: 7, 16, 22, **7**: 17
Agena (rocket) **6**: 41
air resistance **6**: 6, 7, 22
Albor Tholus (Mars volcano) **4**: 41
Aldrin, Jr., E. "Buzz" **3**: 44, **6**: 50
Alpha Regio (Venus) **4**: 25
aluminum, Earth **3**: 39
Amalthea (Jupiter moon) **5**: 34, 35
amino acids **4**: 54
ammonia:
   gas giants **5**: 6
   rocky planets **5**: 6
   Saturn **5**: 39
   Uranus **5**: 49
   See also ammonia-ice
ammonia-ice:
   comets **4**: 54
   Jupiter **5**: 14, 15
   Neptune **5**: 6
   Saturn **5**: 39, 40
   Uranus **5**: 6
Andromeda Galaxy (M31) **1**: 11, 12, 40–41, 50
antenna (pl. antennae) **6**: 31, 40, 52, 53, 54, 55, **7**: 9, 7, 26, 33, 38, **8**: 8, 49, 51, 57
anticyclones **5**: 10, 14, **8**: 20
apogee **3**: 8, 14
Apollo (Moon mission) **3**: 2, 4–5, 16–17, 44–45, 47, 48–49, 51, 52, 54, **6**: 2, 6, 10, 20, 42–51, **7**: 4, 17, 33, 34, 35, 44. *See also* Saturn (launcher)

Apollo applications program.
   See Skylab (space station)
Aqua (satellite) **8**: 14–15, 42
argon:
   Earth **3**: 22
   Moon **3**: 44
Ariane (launcher) **6**: 20, **8**: 10
Ariel (Uranus moon) **5**: 50, 51
Aristarchus (Greek thinker) **1**: 6
Aristarchus (Moon crater) **3**: 50
Armstrong, Neil **6**: 49, 50
Arsia Mons (Mars volcano) **4**: 41
Ascraeus Mons (Mars volcano) **4**: 41
ash:
   Earth **3**: 32, 39, 41
   Mercury **4**: 19
   See also lava
asteroid belt **2**: 47, **4**: 48, 49–50, 57, **6**: 54. *See also* Kirkwood gaps
asteroids **1**: 5, **2**: 4, 46, 51, **4**: 5, 48–52
   collisions with **2**: 51, 53, **4**: 9, 10, 11, 50, 51, 52, **5**: 21
   comet origin **4**: 48
   composition **2**: 51, **4**: 10, 51
   description **4**: 10, 48
   formation **2**: 54, **4**: 11, 50, 51
   irregular **4**: 11, 51
   Mars' moons **4**: 42
   mass **2**: 49, **4**: 51
   meteorite from **4**: 6
   numbers of **4**: 49
   orbits **2**: 46, 47, 51, **4**: 10, 48, 49, 50–51, 52, **6**: 57
   rotation **4**: 51
   See also asteroid belt; Astraea; Ceres; Eros; Gaspra; Hygiea; Ida; Juno; Pallas; Vesta
asthenosphere:
   Earth **3**: 31, 34
   Moon **3**: 55
Astraea (asteroid) **4**: 48
astronauts:
   Apollo (Moon mission) **3**: 44, 45, **6**: 43, 44–45, 46, 47, 49, 50, 51
   endurance **7**: 46
   exercise **7**: 33, 35, 38, 39, 54
   Gemini **6**: 38–39, 41
   gravity, effect of **6**: 41, **7**: 9, 10, 22, **8**: 30
   living space **7**: 23, 33, 34, 39, 41, 45, 50, 53–54
   showering **7**: 34, 35
   space, effect of **7**: 53, 54, 55
   weight **3**: 42
   weightlessness **6**: 16, **7**: 42
   See also cosmonauts; extravehicular activity; manned spaceflight
astronomical unit (AU), definition **1**: 10
astronomy **1**: 42, **7**: 56–57, **8**: 15, 54
astrophysics laboratory (Mir) **7**: 38, 41
asynchronous orbit **8**: 13
Atlantis (Space Shuttle) **7**: 2, 23, 36–37, 40–41
Atlas (launcher) **6**: 19, 35
Atlas (Saturn moon) **5**: 40, 41, 45
atomic mass units **2**: 9
atomic structure **7**: 54
atomic weapons **6**: 18. *See also* ballistic missiles
atomic weight **1**: 14
atoms **1**: 6, 14, 16, 17, 23, 26, 30, 54, 56, **2**: 8, 9, 12, 16, 22, 40, 45, 56, 57, **3**: 24, **4**: 54, **5**: 24, **7**: 57. *See also* atomic structure
Aurora Australis **3**: 21

Aurora Borealis **3**: 21
auroras:
   Earth **2**: 15, 38, 44–45, **3**: 19, 20–21, 24
   Ganymede **5**: 32
   Jupiter **5**: 16
   Uranus **5**: 50

B

ballistic missiles (ICBMs) **6**: 16, 17, 18–19
Beta Regio (Venus) **4**: 25
Betelgeuse (star) **2**: 13, 14
Big Bang **1**: 54, 55, 56, **7**: 55, **8**: 57
binary stars **1**: 18, 26, 30, 36
black dwarf stars **1**: 21, 29, **2**: 11
black holes **1**: 2, 5, 33, 35–39, 45, **8**: 56, 57
blue giants **1**: 14, 17, 22–23, 36, 51, **2**: 12, **8**: 54
Bok globules **1**: 14
booster pods (Russian launchers) **6**: 18, 19
boosters. *See under* Space Shuttle
Borrelly (comet) **4**: 52
bow shock **2**: 40, 41, 42
Butterfly Nebula **1**: 26, **2**: 10

C

calcium:
   stars **1**: 30
   Sun **2**: 16
   universe **1**: 14, 16
Callisto (Galilean satellite) **2**: 55, **5**: 8, 18–19, 30, 33–34
Caloris Basin (Mercury) **4**: 17, 18
Calypso (Saturn moon) **5**: 40, 41
Canada **2**: 45, **3**: 35, **7**: 45, 49, **8**: 21, 25, 42
Candor Chasm (Mars) **4**: 39
carbon:
   Callisto **5**: 34
   comets **4**: 54
   Earth **3**: 5, 16
   interplanetary dust **4**: 10
   Space Shuttle tiles **7**: 28
   stars **1**: 19, 26, 29, **2**: 11
   Sun **2**: 11, 17
   Titan **5**: 42
carbon dioxide:
   comets **4**: 53
   Earth **3**: 22, 27, 30, **4**: 36
   Mars **4**: 34. *See also* carbon dioxide ice
   Venus **4**: 23, 24, 25
carbon dioxide ice (dry ice), Mars **4**: 31, 32, 33, 34
carbon monoxide, comets **4**: 53
Cassini, Giovanni **3**: 9
Cassini division **5**: 44, 46
Cassini-Huygens (probe) **5**: 37, **6**: 21, 56, 57
Cat's Eye Nebula **1**: 26–27
Centaur (launcher) **6**: 21, 56
center of gravity:
   binary stars **1**: 18
   Earth and Moon **3**: 8, 54
centrifugal forces:
   Earth **3**: 10, 18, **4**: 22
   Moon **3**: 10
   Neptune **5**: 53
   satellites **8**: 10
   Saturn **5**: 39
   science fiction **7**: 42
   spacecraft **7**: 10, 42
   Uranus **5**: 49
   Venus **4**: 22
ceramic tiles, heat protection on Space Shuttle **7**: 27–28
Ceres (asteroid) **4**: 10, 48, 51
Cernan, Gene **6**: 51
Challenger (Space Shuttle) **7**: 23, 26–27
Chamberlin, Chrowder **2**: 53
Charon (Pluto moon) **4**: 44, 46–47, 57

Chasma Borealis (Mars) **4**: 32
chromosphere **2**: 18, 22, 32, 33, 35
circumference:
   Earth **3**: 18
   Mars **4**: 41
COBE (satellite) **8**: 57
Columbia (Space Shuttle) **7**: 17, 23, 26
coma (comet) **2**: 51, **4**: 52, 53, 55
combustion chamber. *See* engines
Comet Borrelly **4**: 52
comets **1**: 5, **2**: 4, 46, 51–52, **4**: 5, 6, 8, 10, 47, 48, 53–55
   atmosphere **4**: 53
   collisions with **2**: 52, 53, **5**: 21
   coma **2**: 51, **4**: 52, 53, 55
   Comet Borrelly **4**: 52
   composition **2**: 51, **4**: 8, 10, 53, 54, **5**: 34
   gravitational field **4**: 54
   Halley's Comet **4**: 54, 55
   lifespan **4**: 55
   mass **2**: 14, 49
   nucleus **2**: 51, **4**: 52, 53, 54
   orbits **2**: 51, 52, **4**: 53, 54, 57
   radiation **4**: 53
   Shoemaker-Levy 9 **4**: 53
   tails **2**: 38, 51, **4**: 53, 54
   word origin **4**: 53
   See also Kuiper belt; Oort cloud; *and under* ammonia-ice; carbon; carbon dioxide; carbon monoxide; dust; hydrogen; ice; water; methane; methane-ice; nitrogen; oxygen; plasma; rock; sulfur; water vapor
complementary color **5**: 49, 53
condensation:
   dust formation **1**: 26
   planet and moon formation **2**: 49, 54, **4**: 8, **5**: 4, 6, 17, 18
   water vapor **3**: 23, 29, **4**: 8
conduction **3**: 23, 55
constellations **1**: 8–9
   Andromeda **1**: 9, 23, 40, 41
   Aquila **1**: 9, 30, **6**: 54
   Circinus **1**: 8, 37
   Crux (Southern Cross) **1**: 8
   Draco **1**: 8, 9, 26–27, 47
   Gemini **1**: 8, 28
   Hydra **1**: 8, 35
   Lyra **1**: 9, 12
   Monoceros **1**: 8, 26
   Orion **1**: 8, 9, 13, 17
   Pegasus **1**: 9, 12
   Sagittarius **1**: 9, 11, 17, 18, 25
   Serpens **1**: 8, 16
   Taurus **1**: 9, 13, **6**: 54
   Ursa Major (Great Bear) **1**: 8
   Virgo **1**: 8, 39
   Vulpecula **1**: 9, 13
continental drift **3**: 36–37
convection/convection currents:
   Earth **3**: 23, 29, 32–33, 36
   Ganymede **5**: 32
   Jupiter **5**: 12
   Mars **3**: 36
   Sun **2**: 22, 23, 26, 28
   Venus **4**: 29
convective zone **2**: 18–19, 22, 23
Cooper, Jr., Gordon **6**: 35, 37
Copernicus **1**: 6, **2**: 48
Cordelia (Uranus moon) **5**: 51
corona (pl. coronae) (star) **2**: 19, 22, 35–37, 38, 41, 53, **3**: 15
corona (pl. coronae) (Venus) **4**: 27
coronal loops **2**: 35, 36–37. *See also* prominences
coronal mass ejections **2**: 4–5, 22–23, 35, 44
Cosmic Background Explorer (COBE) **8**: 57
Cosmological Principle **1**: 42, 49
cosmonauts **6**: 31, 32–33, 35, 39, **7**: 30, 32, 39, 41, 46, **8**: 30

65

Cosmonaut Training Center **6:** 33
cosmos **1:** 4, 5, 38, 42, 50, **3:** 5
Crab Nebula **1:** 12, 13, 21, 32–33, 34
Crater Mound. *See* Meteor Crater
craters:
    Callisto **5:** 33, 34
    Dione **5:** 42
    Earth **4:** 9, 52, 57
    Ganymede **5:** 30, 32
    Mars **4:** 38, 40, 41
    Mercury **4:** 9, 14, 16, 18, 19
    Mimas (Herschel Crater) **5:** 42
    Moon **2:** 51, **3:** 42, 45, 46, 47, 50–51, 52, 55, 56, 57, **4:** 9
    Neptune's irregular moons **5:** 56
    Oberon **5:** 51
    Phobos **4:** 42
    Pluto **4:** 45
    Tethys **5:** 42
    Umbriel **5:** 51
    Venus **4:** 25, 28, 29
    See also Aristarchus (Moon crater); Cunitz Crater (Venus); Daedalus (Moon crater); Leonov Crater (Moon); Meteor Crater; rays, crater
crescent Moon **3:** 13. *See also* phases
Crux (constellation) **1:** 8
Cunitz, Maria **4:** 28
Cunitz Crater (Venus) **4:** 28
cyclones **3:** 26, **5:** 10, **8:** 23
    tropical cyclones **8:** 22
    See also hurricanes

D

Daedalus (Moon crater) **3:** 51
dark matter **1:** 5, 40
deep space network (NASA) **2:** 20
Deimos (Mars moon) **4:** 42
Delta (launcher) **8:** 12–13
deorbit **7:** 32, 35
depressions (weather) **8:** 20, 21, 22, 25
deserts:
    Earth **3:** 36, 40, **4:** 38, **8:** 38, 43, 46
    Mars **4:** 38
DG Tau (star) **1:** 48
"diamond ring" (Sun) **3:** 14–15
diffraction **3:** 44
dinosaurs, extinction of **2:** 51, **4:** 52
Dione (Saturn moon) **5:** 36, 40, 42, 43, 45
Discoverer (satellite) **6:** 26, 27
Discovery (Space Shuttle) **7:** 16, 23, 51
docking, spacecraft **6:** 2, 3, 38, 40, 41, 44, 45, 46, 49, **7:** 13, 30, 32, 33, 34, 36–37, 38, 41, 44, 46, 52, 53
dog (Laika), first living creature in space **6:** 23
Doppler effect **6:** 23
"double planet" **3:** 18, **4:** 46
drag **7:** 9
Dreyer, John Ludvig Emil **1:** 12
dry ice. *See* carbon dioxide ice
Dumbbell Nebula **1:** 13
dust **1:** 4, 14, 17, 20, 26, 29, 36, 40, 45, **2:** 8, **4:** 54, **8:** 54
    comets **2:** 52, 55, **4:** 52, 53
    Earth **2:** 53, **8:** 4–5
    Mars **4:** 34
    Moon **3:** 44
    solar system origin **2:** 53, 54
    See also gas and dust clouds; interplanetary dust; interstellar matter
dust storms, Mars **4:** 33, 34, 35, 36–37
dwarf stars **1:** 19, 22, **2:** 6, 10, 12.
    See also black dwarf stars; white dwarf stars

E

"Eagle." *See* lunar module
Eagle Nebula **1:** 16
Earth **2:** 46, 47, 48, 50, 55, **3:** 16–41, **4:** 4, 5
    atmosphere **2:** 9, 23, 26, 41, **3:** 5, 6–7, 22–27, 30, 36, 39, 45, 52, **6:** 8, 11, 23, 25, **7:** 6–7, 41, 56, **8:** 8, 20–27, 41. *See also* exosphere; ionosphere; mesosphere; stratosphere; thermosphere; troposphere
    atmospheric pressure **4:** 23
    auroras **2:** 15, 38, 44–45, **3:** 19, 20–21, 24
    "greenhouse" effect **4:** 36
    sky, color of **3:** 44
    storms **3:** 26, **8:** 23, 25
    weather patterns **8:** 19, 20–27
    winds **3:** 29, 40, **8:** 43
    axis **2:** 14, **3:** 8, 11, 16, 18
    centrifugal forces **3:** 10, 18, **4:** 22
    circumference **3:** 18
    death **2:** 10
    density **3:** 19
    energy **3:** 29, 32, 39
    formation **3:** 30, 39
    gravity/gravitational field **3:** 3, 10, 18, **6:** 8, 51, **7:** 10, **8:** 18–19
    Earth and Moon center of gravity **3:** 8, 54
    escaping from **6:** 8, 9, 22, 33, **7:** 9, 10, **8:** 10, 13
    formation **3:** 30
    simulation **7:** 42, 56
    heat **3:** 22, 23, 27, 29, 30, 32, 39
    inside:
        asthenosphere **3:** 31, 34
        convection currents **3:** 23, 29, 32–33, 36
        core **3:** 19, 31, 38, 39
        crust **3:** 19, 30–35, 36–37, 38, 39, 40, 41, **4:** 51
        lithosphere **3:** 30, 31
        mantle **3:** 30, 31, 34, 38, 39, 40
        plates **3:** 30–37, 39, 41
    magnetism/magnetic field **2:** 42, 45, **3:** 19–21, 24, 38, **5:** 16, **6:** 26, **7:** 56, **8:** 56. *See also* magnetosphere
    magnetosphere **2:** 38, 42, 44, 45, **3:** 20–21, 24
    Mars, Earth seen from **3:** 6
    mass **2:** 53, **3:** 19, 39
    Moon. *see* Moon
    orbit **2:** 47, **3:** 8, 10, 11, 16, 18, **4:** 7, 22, **6:** 56, 57
    pressure **2:** 16, **3:** 38, 39, 42
    radiation **4:** 36, **8:** 26, 57. *See also* Van Allen belts
    radioactive decay **3:** 32, 39
    radio waves **2:** 41, **3:** 22, **6:** 23
    radius **3:** 19
    rotation **2:** 14, **7:** 13
    seasons **3:** 10–11, **8:** 19
    shape and size **3:** 5, 16, 18–19, **4:** 5
    surface area **3:** 19
    surface features. *See* glaciers; *and under* ash; craters; deserts; erosion; faults/fractures; flooding; ice caps; lava/lava flows; magma; mountains/mountain ranges; oceans; rifts/rift valleys; sand dunes; "soils"; volcanoes/volcanic activity
    temperature **3:** 27, 28, 39, **8:** 27
    tilt **3:** 8, 11, 16
    See also aluminum; earthquakes; surveying (and mapping) Earth; *and under* argon; carbon;

Earth (*continued...*)
    carbon dioxide; dust; helium; hydrogen; ice, water; iron; light; magnesium; meteorites; meteoroids; nitrogen; oxygen; phases; reflect, ability to; rock; silicate minerals; silicon; snow; sulfur; tides; ultraviolet radiation; water; water vapor
earthquakes **2:** 16, 32, **3:** 32, 33, 38, **8:** 46
eccentric orbits:
    Moon **3:** 9
    Nereid **5:** 56
    Pluto **4:** 44
Echo (satellite) **6:** 28, **8:** 8
eclipses **2:** 14, 32, 33, 47, **3:** 9, 14–15
ecliptic plane **2:** 47
EGGs (evaporating gaseous globules) **1:** 16
Einstein, Albert **1:** 46, 48, **2:** 9
electronics **6:** 27, **7:** 6, 18, **8:** 19, 20
electrons **1:** 34, 36, 39, **2:** 8, 9, 32, 41, 42, 45, **3:** 19
elements. *See individual entries*
El Niño **8:** 20
Elysium Mons (Mars volcano) **4:** 41
Enceladus (Saturn moon) **5:** 6–7, 36, 40, 42, 44
Encke gap (Saturn rings) **5:** 44, 46
Endeavour (Space Shuttle) **7:** 23, 51
energy:
    Big Bang **1:** 54
    Earth **3:** 29, 32, 39
    galaxies **1:** 40, **2:** 8
    Io **5:** 22
    Jupiter **5:** 8, 15
    meteor impact **4:** 57
    quasars **1:** 35
    relativity, theory of **1:** 46, 48, **2:** 9
    stars **1:** 21, 23, 26, 30, 31, 35, 45
    Sun **2:** 8–9, 16, 21, 22, 28, 31, 32, 33, 38, 46, **3:** 22, 28, **8:** 26
    universe **1:** 4
    Uranus **5:** 50
engines (launcher and spacecraft) **6:** 10, 11–13, 16, 19, **8:** 14
    catalysts **7:** 18, 20
    combustion chamber **6:** 12, 13, **7:** 20
    cryogenic **6:** 20
    Goddard's double acting **6:** 14
    gravity, overcoming **7:** 9, 10, 13
    hypergolic **6:** 11, **7:** 20
    lunar module **6:** 46, 47
    Mir **7:** 38, 39
    oxidizers **6:** 11, 12, 13, 16
    reaction **6:** 6, 12–13, **7:** 20
    Skylab **7:** 34
    Space Shuttle **6:** 13, **7:** 14, 15, 18, 20, 21, 23, 27
    inertial upper stage **7:** 25
    Titan **6:** 56
    See also fuel; propulsion systems; takeoffs; thrusters
Enterprise (Space Shuttle) **7:** 17, 23
Eros (asteroid) **4:** 50
erosion:
    Callisto **5:** 33
    Earth **3:** 23, 28, 29, 34, 38, 40, 41, **8:** 44
    Mars **4:** 40
    Moon **3:** 49, 50
ERTS-1 (satellite) **8:** 28–29, 34.
    See also Landsat
ESA (European Space Agency) **6:** 20, **7:** 49, **8:** 7, 10
Eskimo Nebula **1:** 28, **2:** 11
Eta Carinae (nebula) **1:** 10
EUE (Extreme Ultraviolet Explorer) **8:** 57
Europa (Galilean satellite) **2:** 50, 55, **5:** 8, 16, 18–19, 26, 28–29, 34

Europe **3:** 34, 37, **6:** 20, 28, **7:** 45
European Space Agency. *See* ESA
European space initiatives. *See* Ariane; ESA; International Space Station; Meteosat
EVA. *See* extravehicular activity
evaporation:
    dry ice **4:** 34
    star formation **1:** 16
    Sun death **2:** 10
    water cycle **3:** 28, 29
exosphere **3:** 22
expanding universe **1:** 42, 53–57
experiments in space **6:** 33, 50, **7:** 24, 26, 41, 49, 54–57.
    See also laboratories in space
Explorer (satellite) **6:** 17, 24–25, **8:** 8, 28, 56
external tank. *See under* Space Shuttle
extinction, dinosaur **2:** 51, **4:** 52
extravehicular activity (EVA) (spacewalking) **6:** 38–39, 41, 50, 51, **7:** 36, 41, 44, 46–47
Extreme Ultraviolet Explorer (EUE) **8:** 57

F

Faith 7 (Mercury mission) **6:** 37
false-color images, explained **8:** 25, 34
faults/fractures:
    Dione **5:** 42
    Earth **3:** 33, **8:** 37, 46
    Mercury **4:** 17, 18
    Venus **4:** 26–27, 28
    See also corona (Venus); nova (Venus); rifts/rift valleys
filters **2:** 32, 33. *See also* multispectral images
flares **2:** 19, 34, 32–33, 35, 45
flight paths **6:** 56, 57, **7:** 10–11, 20. *See also* orbits; trajectory
flooding:
    Earth **8:** 23, 25, 30, 38
    Mars **4:** 31
fractures. *See* faults/fractures
Freedom 7 (Mercury mission) **6:** 34
friction **3:** 9, 6, **7:** 10, 27. *See also* air resistance
Friendship 7 (Mercury mission) **6:** 34, 35, 36, 37
fuel, spacecraft **6:** 7, 8, 9, 10, 11, 12, 13, 14, 16, 19, 20, 31, 44, 46, 52, 56, **7:** 14, 15, 18, 19, 20. *See also* propellants
full Moon **3:** 9, 12, 13, 47. *See also* phases
fusion **1:** 18, 20, **2:** 8, 9, 10, 21

G

G2V star (Sun) **2:** 12
Gagarin, Yuri **6:** 32–33
galaxies **1:** 4, 40–45, **2:** 6, **8:** 56, 57
    cataloguing **1:** 12
    definition **1:** 40
    formation **1:** 42, 54, 55, 56
    gravity/gravitational field **1:** 42, 47, 56
    mass **1:** 24, 40
    size **1:** 11
    star rotation **1:** 42
    types **1:** 42–43, 45
        elliptical **1:** 19, 23, 39, 42, 43, 45, 46, 47, **5:** 6
        irregular **1:** 18, 42, 43, 45, 51
        radio **1:** 45
        spiral **1:** 19, 30, 40, 42, 43, 45, 46–47, 50, 52–53, **2:** 6
    See also Andromeda Galaxy; Big Bang; Hubble's Law; Local Group; Magellanic Clouds; Milky Way Galaxy; *and under* energy; helium; hydrogen; light; reflect, ability to; X-rays

Galilean satellites **5:** 18, 19–34.
    See also Callisto; Europa;
    Ganymede; Io
Galilei, Galileo **1:** 6, **2:** 48, **5:** 18
Galileo (probe) **3:** 42, **4:** 20, **5:** 9,
    10, 12–13, 18, 21, 23, **6:** 56, 57
gamma rays **2:** 21, 33
Ganymede (Galilean satellite)
    2: 55, **5:** 6, 8, 16, 18–19, 26,
    30–32, 34
gas, interstellar **1:** 4, 14, 16, 17,
    19, 20, 26, 28, 29, 36, 37, 39,
    40, 56, **2:** 8, **8:** 54. *See also* gas
    and dust clouds; ionized gas
gas and dust clouds **1:** 14, 17–18,
    19, 20, 23, 26–27, 30–31, 33,
    42, **2:** 8, 53, 54, **3:** 56. *See also*
    nebula
gas giant moons **2:** 50, 57, **5:** 6–7.
    See also Galilean satellites;
    and under Jupiter; Neptune;
    Saturn; Uranus
gas giants **2:** 50, **5:** 4 *and*
    throughout
    formation **2:** 56–57, **5:** 4, 6, 17,
        18, 48, 49, 50
    ring formation **5:** 7, 34, 35,
        45, 46
    See also Jupiter; Neptune;
        Saturn; Uranus
gas plumes:
    Io **5:** 23, 25
    Neptune (geysers) **5:** 56
Gaspra (asteroid) **4:** 10, **6:** 57
geiger tubes **6:** 24
Gemini (spacecraft) **6:** 2, 3, 38–41
George C. Marshall Space Flight
    Center **6:** 17
geostationary orbit **7:** 5, 25, **8:** 13
geostationary satellites **6:** 28, **8:** 7,
    14, 22, 26. *See also* global
    positioning system
geosynchronous orbit **7:** 10, 12–13,
    8: 13, 14–15, 19, 23
geosynchronous satellites **7:** 12, 13
German Rocket Society **6:** 16
Germany **6:** 14, 16
geysers, on Triton **5:** 56
gibbous Moon **3:** 12, 13. *See also*
    phases
gimbals **6:** 12, 13, 16, **7:** 20, 25
glaciers **2:** 32, **3:** 28, **8:** 30, 31, 38
glass:
    fiber **7:** 28
    Moon rock **3:** 52
    "stardust" **4:** 10
Glenn, John **6:** 34, 35
global positioning system (GPS)
    6: 23, **7:** 5, **8:** 4, 7, 13, 14
Goddard, Robert **6:** 14–16, 17
GOES (satellite) **8:** 17, 23–26
Grace (satellite) **8:** 18–19
granulation **2:** 24, 26, 27, 28
gravity/gravitational field:
    "assists" (slingshot) **4:** 14, 54,
        6: 52, 55, 56
    asteroids **4:** 10, 11, 51
    astronauts **6:** 41, **7:** 9, 10, 22,
        8: 30
    Big Bang **1:** 56
    black holes **1:** 33, 36
    Callisto **5:** 34
    comets: **4:** 54
    cosmos **1:** 5
    craters **3:** 51
    Earth **3:** 8, 10, 18, **6:** 8, 51, **7:** 10,
        8: 18–19
    escaping from **6:** 6, 8, 9, 22,
        33, **7:** 9, 10, **8:** 10, 13
    formation **3:** 30
    simulation **7:** 42, 56
Europa **5:** 34
experiments (ISS) **7:** 54–57
galaxies **1:** 42, 47, 56
Ganymede **5:** 34
Io **5:** 22, 34

gravity (*continued...*)
    Jupiter **4:** 49, **5:** 6, 8, 17, 22, 34
    light, effect on **1:** 46, 47
    Mars and its moons **4:** 30, 33, 42
    Mercury **4:** 12, 15
    Moon **3:** 8, 10, 42, 44, 51, 54
    Neptune **5:** 5, 56
    pull (g) **7:** 9, 22
    Saturn **5:** 6, 39, 45, 46
    solar system origin **2:** 53, 54, 57
    sphere-making **4:** 11, **5:** 34, 56
    stars **1:** 18, 19, 20, 23, 24, 26, 29,
        30, 31, 42, 56, **2:** 8, 9
    Sun **2:** 8, 9, 14, 16, 46, **3:** 10,
        8: 56
    understanding **7:** 57
    universe, all objects in **3:** 18
    Uranus **5:** 49, 50
    variance **8:** 18–19
    See also center of gravity; laws
        of motion; Newton, Sir
        Isaac; theory of relativity
Great Bear **1:** 8
Great Dark Spot (Neptune) **5:** 54
Great Red Spot (Jupiter) **5:** 4–5,
    10, 14, 15
"greenhouse" effect:
    Earth **3:** 36
    Venus **4:** 24–25
ground station **6:** 28, 31, **7:** 6,
    8: 8, 9, 13, 14
Gula Mons (Venus) **4:** 28
gyroscopes **6:** 12, 16

H
half Moon **3:** 12, 13. *See also*
    phases
Halley, Sir Edmund **4:** 55
Halley's Comet **4:** 54, 55
heat shields **5:** 12, **6:** 26, 33, 40, 44,
    46, 51, **7:** 27, **8:** 57
Hecates Tholus (Mars volcano)
Helene (Saturn moon) **5:** 40, 41
heliopause **2:** 40, 41
helioseismology **2:** 16, 32
heliosphere (Sun) **2:** 40–41, 44, 45
heliotail **2:** 40
helium:
    Big Bang **1:** 54
    Earth **3:** 22
    galaxies **1:** 42
    gas giants **2:** 50, **5:** 6
    Jupiter **5:** 6, 15, 17
    Mercury **4:** 14, 16
    Moon **3:** 44
    Neptune **5:** 6, 53
    Saturn **5:** 6, 39
    space **2:** 56–57
    stars **1:** 14, 21, 23, 24, 26, 28,
        29, 42
    Sun **2:** 8–9, 11, 17, 21, 22, 40
    universe **1:** 14
    Uranus **5:** 6, 48, 49
    word origin **2:** 17
Hellas Basin (Mars) **4:** 41
Heracleides (Greek thinker) **1:** 6
Herschel, William and John **1:** 7, 12
Herschel Crater (Mimas) **5:** 42
Hertzsprung, Ejnor **2:** 10
Hertzsprung-Russell diagram **2:** 11
Hubble, Edwin Powell **1:** 42–43,
    45, 53
Hubble Space Telescope **7:** 2, 5, 17,
    56, **8:** 2, 54, 56, 57
    how Hubble "sees" **1:** 7
    images taken by **1:** 7 *and*
        throughout, **4:** 30, 31,
        32–33, 44, 45; **5:** 16,
        8: 54, 55
Hubble's Law **1:** 42, 53
Hubble-V (gas cloud) **1:** 18
hurricanes **3:** 26, **7:** 57, **8:** 20, 22,
    23, 25, 26
Huygens gap (Saturn rings) **5:** 46
Huygens (probe) **6:** 21, 57. *See also*
    Cassini-Huygens

hydrazine **6:** 11, **7:** 20
hydrogen:
    Big Bang **1:** 54
    comets **4:** 54
    Earth **3:** 22
    galaxies **1:** 45
    gas giants **2:** 50, 56–57, **5:** 6
    interstellar wind **2:** 40
    Jupiter **5:** 6, 15, 17
    Lagoon Nebula **1:** 11
    Mercury **4:** 16
    Moon **3:** 44
    Neptune **5:** 6, 53
    rocky planets **2:** 56–57
    Saturn **5:** 6, 38, 39
    spacecraft fuel (liquid
        hydrogen) **6:** 11, 56, **7:** 14,
        15, 18, 20, 21
    stars **1:** 14, 16, 17, 20, 21, 23, 24,
        26, 28, 29, 42, 45
    Sun **2:** 8–9, 10, 12, 17, 21, 22, 40
    universe **1:** 14
    Uranus **5:** 6, 48, 49
Hygiea (asteroid) **4:** 51
Hyperion (Saturn moon) **5:** 41, 45

I
Iapetus (Saturn moon) **5:** 41, 42,
    43, 45
ICBMs (intercontinental ballistic
    missiles). *See* ballistic missiles
ice, water
    Callisto **5:** 30, 33, 34
    Charon **4:** 46
    comets **2:** 46, 51, **4:** 54
    condensation **4:** 8
    Earth **3:** 6, 22, 28, **8:** 25, 30,
        38, 41
    Europa **2:** 50, **5:** 19, 28–29
    Ganymede **5:** 30, 32
    gas giant rings **5:** 6
    gas giants **5:** 4
    Iapetus **5:** 42
    Kuiper belt **4:** 47
    Mars **4:** 31, 34, 35
    Neptune **5:** 6, 55
    Pluto **4:** 8, 44
    Saturn/Saturn's rings **5:** 38,
        40, 45
    solar system origin **2:** 56–57
    Tethys **5:** 42
    Triton **5:** 56
    Uranus/Uranus' moons **5:** 6, 49,
        50, 51
    See also ammonia-ice; carbon
        dioxide ice (dry ice);
        ice caps; methane-ice;
        nitrogen-ice
ice caps:
    Earth **3:** 28, **4:** 32
    Mars **4:** 31, 32–33, 34, 36
    Mercury **4:** 17
    See also polar caps
Ida (asteroid) **4:** 10, **6:** 57
inertial upper stage (IUS) **7:** 25
inflation **1:** 54, 55. *See also*
    expanding universe
infrared:
    false-color images **8:** 34–35, 37
    satellites **6:** 27, **8:** 8, 22, 24, 25,
        26, 44, 56, 57
    stars **1:** 17, 18, **8:** 57
    Sun, light energy from **2:** 9
    thermal imaging **8:** 41
Infrared Astronomical Satellite
    (IRAS) **8:** 57
inner planets. *See* Mercury; Venus;
    Earth; Mars
INTELSAT VI (satellite) **8:** 10–11
interferogram **8:** 51
interferometry **8:** 51
International Space Station (ISS)
    7: 16, 41, 42–57, **8:** 4
    crew **7:** 52, 53–54
    modules:
        Destiny lab **7:** 45, 52

International Space Station (ISS)
    (*continued...*)
    emergency crew vehicle
        (X-38) **7:** 44, 53
    laboratory **7:** 49, 52, 54–57
    Leonardo **7:** 45, 52
    Pirs **7:** 44, 52, 53
    Raffaello **7:** 52
    remote manipulator system
        (RMS) **7:** 45, 46–47,
        49, 52
    solar panels **7:** 42, 45
    trusses **7:** 45, 52, 53
    Unity **7:** 42–43, 45, 50, 51, 52
    Zarya **7:** 42–43, 45, 50, 51, 52
    Zvezda **7:** 44, 52
    orbit **7:** 46
    Freedom, former name **7:** 44
    living space **7:** 53–54
    Mir, learning from **7:** 46
    pictures taken from **8:** 30, 31
interplanetary dust **4:** 10, 11. *See*
    also dust; interstellar matter;
    micrometeorites
interstellar gas. *See* gas, interstellar
interstellar matter **1:** 14, 23, 40,
    2: 40, 41. *See also* dust; gas,
    interstellar; interplanetary
    dust; micrometeorites
Io (Galilean satellite) **2:** 50, 55,
    5: 8, 16, 18, 19, 20–26, 34
ionization **2:** 8
ionized gas **1:** 17, **2:** 26, 33, 35, 46.
    See also plasma
ionized particles **3:** 24, **5:** 26.
    See also auroras; plasma;
    prominences; spicules
ionosphere **2:** 41, 45, **3:** 21, 22, 24,
    6: 23
IRAS (Infrared Astronomical
    Satellite) **8:** 57
iron:
    Earth **3:** 19, 39
    Europa **5:** 29
    Io **5:** 26
    Mars **4:** 31, 43
    Mercury **4:** 12, 17, 19
    meteorites **4:** 6, 57
    Moon **3:** 52, 56
    oxide (catalyst) **7:** 18
    rocky planets **4:** 8
    stars **1:** 19, 30, 31
    Sun **2:** 17
isotopes, radioactive.
    See radioisotopes
Italy **7:** 45, 52, **8:** 31

J
Janus (Saturn moon) **5:** 40, 44, 46
Japan **7:** 45, 49, **8:** 37
Jeans, Sir James **2:** 53
Jeffreys, Sir Harold **2:** 53
Jovian planets **3:** 6, **4:** 8, **5:** 4, 6.
    See also gas giants
Juno (asteroid) **4:** 48
Jupiter (launcher) **6:** 19, 24
Jupiter (planet) **2:** 46, 47, 49, 50,
    55, **4:** 8, 48, 49, 57, **5:** 2, 4, 5,
    8–18, **6:** 52, 54, 55, 56, 57
    atmosphere **5:** 9, 10–15
    anticyclones **5:** 10, 14
    atmospheric pressure **5:** 12
    auroras **5:** 16
    clouds **5:** 10–11, 14–15
    convection currents **5:** 12
    storms **5:** 14
    weather patterns **5:** 10–15
    winds **5:** 14
    axis **5:** 9
    composition **5:** 4, 6, 8, 9, 15,
        16–17
    density **5:** 8
    direction **5:** 4, 8
    formation **2:** 56–57, **5:** 4–5,
        17, 18
    gravity **4:** 49, **5:** 6, 8, 17, 22, 34

67

Jupiter (continued...)
    Great Red Spot 5: 4–5, 10, 14, 15
    heat 5: 8, 12, 16, 17
    magnetism/magnetic field 5: 8, 15–16, 26, 6: 54, 55
    magnetosphere 5: 15, 32
    mass 2: 49, 5: 8, 17
    moons 5: 6–7, 8, 9, 15, 16, 18–35. See also Galilean satellites
    orbit 2: 47, 5: 4, 6: 56, 57
    pressure 5: 17
    probes. See Galileo; Pioneer; Voyager
    radiation 5: 15
    radio waves 5: 8, 15
    rings 5: 6, 7, 8, 34, 35
    rotation 5: 9, 35
    shape and size 5: 5, 8
    star, potential to be 5: 8, 17
    temperature 5: 11, 12, 16, 17
    See also Shoemaker-Levy 9; and under ammonia-ice; energy; helium; hydrogen; metallic hydrogen; methane; phosphorous; polar caps; rock; sulfur; water

K

Kant, Immanuel 2: 53, 54
"Kaputnik" 6: 24
Keeler gap (Saturn rings) 5: 46
Kennedy, President John F. 6: 17, 38
Kennedy Space Center 6: 4, 19, 7: 20
Kepler, Johannes 1: 6, 2: 48, 3: 9
kerosene 6: 11, 19
kilopasec, definition 1: 11
Kirkwood gaps (asteroid belt) 4: 49
Kuiper belt 2: 51, 52, 4: 47, 57

L

laboratories in space:
    International Space Station (ISS) 7: 49, 52, 54–57
    Mir 7: 38, 41
    Space Shuttle 7: 24, 26
Lagoon Nebula 1: 11
Laika, the dog (first living creature in space) 6: 23
Landsat 8: 14, 34–39, 45
landslides, Mars 4: 41
La Niña 8: 20
Laplace, Pierre-Simon 2: 53
launchers/launch vehicles 6: 4–5, 9–13, 14–21, 24, 38, 43, 52, 56, 7: 34, 46, 8: 10. See also Ariane (launcher); Atlas (launcher); engines; Jupiter (launcher); propellants; propulsion systems; retrorockets; rocketry; rockets; Saturn (launcher); takeoffs; Thor (launcher); thrusters; Titan (launcher); Vanguard (launcher)
lava/lava flows:
    Dione 5: 42
    Earth 3: 29, 32, 38, 39, 41
    Io 5: 21, 22, 23, 25–26
    Mars 4: 40, 43
    Mercury 4: 17, 18, 19
    Moon 3: 42, 46, 47, 48–49, 52, 54, 55, 56
    Venus 4: 25, 26, 27, 28, 29
laws of motion 1: 6, 2: 38, 53, 6: 12, 41
Leonov Crater (Moon) 3: 47
life:
    development 1: 50, 2: 9, 46, 3: 5, 16, 22, 27, 30, 4: 54
    evidence of on other planets/moons 1: 48, 5: 29
    extinction 2: 10, 51, 3: 19, 4: 52
    first human being in space 6: 33

life (continued...)
    first living creature in space 6: 23, 30
    identifying life on Earth from space 3: 6–7
lift/lifting force. See under rockets
liftoffs. See takeoffs
light:
    Big Bang 1: 55, 56
    black hole 1: 36
    Earth 2: 6, 3: 13
        blue sky 3: 44
    galaxies 1: 40, 45, 50, 53
    Ganymede 5: 30
    gravity, effect of 1: 46, 47
    Mars 4: 31, 36
    Moon 3: 13, 47
    Neptune 5: 53
    quasars 1: 35
    rays of 1: 36
    speed of 1: 10
    stars 1: 16, 18, 24, 29, 30, 33, 39, 42, 45
    Sun 2: 6, 9, 11, 16–17, 21, 23, 24, 32, 41, 46, 3: 11, 13, 14–15, 22
        elements within 2: 16–17
    time machine 1: 50, 53
    universe 1: 7
    visible 1: 5, 18, 30, 2: 9, 21, 32, 4: 20, 8: 22, 24, 33, 34, 38.
    See also complementary color; filters; infrared; light-year; photons; radiation; ultraviolet light; X-rays
light-year, definition 1: 10–11
lithosphere:
    Earth 3: 30, 31
    Moon 3: 55
Local Group 1: 50, 2: 7
lunar eclipse 3: 14
lunar module (LM) ("Eagle") 6: 20, 43, 44, 45, 46, 47, 48, 49, 50. See also Apollo (Moon mission)

M

megaparsec, definition 1: 11
Magellan, Ferdinand 1: 51, 4: 25
Magellan (probe) 4: 25
Magellanic Clouds 1: 11, 45, 50, 51, 2: 7, 8: 54
magma:
    Earth 3: 32, 33, 38, 40
    Venus 4: 26
    See also lava/lava flows
magnesium:
    Earth 3: 39
    Sun 2: 17
magnetism/magnetic field:
    black holes 1: 2, 39
    cosmos 1: 5
    Earth 2: 42, 45, 3: 19–21, 24, 38, 5: 16, 6: 26, 7: 56, 8: 56
    Ganymede 5: 32
    Io 5: 15, 16
    Jupiter 5: 8, 15–16, 26, 6: 54, 55
    Mercury 4: 19
    Moon 3: 55
    Neptune 5: 55, 6: 55
    Saturn 5: 37, 38, 6: 55
    Sun/sunspots 2: 16, 28, 32, 33, 35, 37, 41, 42–45, 46
    supernova 1: 34
    Uranus 5: 49, 50, 6: 55
    See also magnetopause; magnetospheres
magnetopause 2: 42
magnetospheres:
    Earth 2: 38, 42, 44, 45, 3: 20–21, 24
    Ganymede 5: 32
    Jupiter 5: 15, 32
    Sun's heliosphere 2: 40–41
    See also magnetopause; magnetotail

magnetotail 3: 21
magnitude, star 2: 13, 8: 56
main-sequence stars 1: 18, 24, 30, 2: 9, 10, 12
man, first in space 6: 32–33
manned spaceflight 3: 6, 6: 6, 9, 24, 26, 30–51, 52, 8: 30. See also Apollo; Gemini; Soyuz; Space Shuttle
mare (pl. maria) 3: 42, 43, 46, 47, 52, 54, 55, 4: 17, 31, 38
Mariner (probe) 4: 12, 14, 15, 16, 6: 52, 53
Mars 3: 42, 46, 47, 48, 50, 55, 4: 2, 4, 5, 6, 30–43, 6: 52
    atmosphere 4: 33–37
        clouds 4: 31, 32, 34, 35
        convection 4: 36
        dust storms 4: 33, 34, 35, 36–37
        weather patterns 4: 34
        winds 4: 33, 36, 38
    axis 4: 32
    circumference 4: 41
    color 4: 30, 31
    density 4: 43
    formation 4: 43
    gravity 4: 30, 33, 42
    heat 4: 36, 43
    inside:
        core 4: 43
        crust 4: 41, 43
        mantle 4: 43
        plates 4: 41
    mass 4: 30
    moons 4: 42
    orbit 2: 47, 4: 7, 31, 32, 50
    pressure 4: 34
    probes. See Mars Global Surveyor; Viking mission
    radiation 4: 43
    radioactive decay 4: 43
    radius 4: 43
    rotation 4: 31, 32
    seasons 4: 32
    shape and size 4: 5, 30
    surface features:
        basins 4: 41
        "canals" 4: 38
        chasm 4: 32, 39
        craters 4: 38, 40, 41
        deserts 4: 38
        erosion 4: 40
        flooding 4: 31
        gullies 4: 40
        highlands 4: 38
        ice caps 4: 31, 32–33, 34, 36
        landslides 4: 41
        lava/lava flows 4: 40, 43
        mare (pl. maria) 4: 38
        plains 4: 40, 41
        polar caps 4: 31, 34
        ridges 4: 38
        rift valleys 4: 41
        sand dunes 4: 36
        "soils" 4: 34, 43
        volcanoes 4: 38, 40, 41, 43
    temperature 4: 34, 36
    tilt 4: 32
    See also noble gases; and under carbon dioxide; carbon dioxide ice (dry ice); dust; ice, water; iron; light; nitrogen; oxygen; reflect, ability to; rock; water; water vapor
Mars Global Surveyor (probe) 4: 30
mascons (Moon) 3: 54, 55
mathematics, laws of 1: 6
matter, conversion into energy by:
    galaxies 1: 40
    Sun 1: 23, 2: 9, 21, 22
Maxwell gap (Saturn rings) 5: 46
Maxwell Montes (Venus) 4: 25, 26
measurements in space 1: 10–11
Melas Chasm (Mars) 4: 39

Mercury (planet) 1: 6, 9, 2: 46, 47, 48, 50, 55, 4: 4, 5, 6, 7, 9, 12–19, 6: 52
    atmosphere 4: 14, 15–17
    axis 4: 15
    death 2: 10
    density 4: 12
    gravity/gravitational field 4: 12, 15
    heat 4: 16
    inside:
        core 4: 12, 19
        crust 4: 12, 16, 18, 19
        mantle 4: 19
    magnetic field 4: 19
    orbit 2: 47, 4: 7, 12, 14, 15, 16
    probes. See Mariner (probe)
    radioactive decay 4: 43
    rotation 4: 15
    shape and size 4: 5, 12, 14, 19
    surface features 4: 17–19
        basins 4: 17, 18
        craters 4: 9, 14, 16, 18, 19
        depressions 4: 18
        faults/fractures 4: 17, 18
        ice caps 4: 17
        lava/lava flows 4: 17, 18, 19
        mare (pl. maria) 4: 17
        mountains 4: 16, 17, 18
        plains 4: 16, 19
        ridges 4: 16, 17, 19
        scarps (escarpments) 4: 16, 18
        "soils" 4: 17, 18
        valleys 4: 16
    temperature 4: 16
    See also under ash; helium; hydrogen; iron; oxygen; phases; potassium; reflect, ability to; sodium; titanium
Mercury (spacecraft) 6: 30, 34–37. See also Faith 7; Freedom 7; Friendship 7
mesosphere 3: 22, 24
messages, from Earth into space 6: 54
Messier, Charles 1: 12
Messier Catalogue 1: 12, 13, 33
metallic hydrogen:
    Jupiter 5: 17
    Saturn 5: 38
Meteor Crater, Arizona 4: 9, 52
meteorites 4: 6, 55–57
    age 4: 11
    analysis 2: 52, 57, 7: 57
    asteroid origin 4: 57
    Big Bang 1: 54
    definition 4: 11, 55
    Earth, collisions 2: 57, 3: 30, 45, 4: 52, 55, 57
    images of 4: 6, 11, 38
    Mars, collisions 4: 57
    Mercury, collisions 4: 18, 19
    Moon, collisions 3: 45, 48–49, 50, 52, 57
    Venus, collisions 4: 29.
    See also meteoroids; meteors; micrometeorites; and under asteroids; iron; nickel; silicate minerals
meteoroids 4: 55–57
    asteroid origin 4: 57
    definition 4: 11, 55
    discoveries 4: 6
    Earth, collisions 4: 10, 11
    gas giant rings 5: 7, 34, 35
    Jupiter's rings 5: 34, 35
    mass 2: 49
    Mercury, bombardment 4: 18
    planet formation 4: 9
    solar system 4: 5, 10, 40
    See also meteorites; meteors; micrometeorites
meteorological satellites.
    See weather satellites
meteorology 8: 20, 24, 25, 7: 12.
    See also weather satellites

meteors **4:** 55–57
  asteroid origin **2:** 51, 52
  climate change **2:** 51, **4:** 52
  definition **4:** 11, 55
  dinosaur extinction **2:** 51, **4:** 52
  meteor craters **2:** 51. *See also*
    Meteor Crater, Arizona
  orbits **4:** 57
  "shooting stars" **4:** 11
  See also meteorites; meteoroids;
    meteor showers;
    micrometeorites
meteor showers **2:** 52, **4:** 57
Meteosat (satellite) **8:** 25
methane:
  comets **4:** 54
  gas giants **5:** 6
  Jupiter **5:** 15
  Neptune **5:** 6, 52, 53
  Pluto **4:** 44, 45, 46
  rocky planets **5:** 6
  Saturn **5:** 39
  Titan **5:** 41
  Uranus **5:** 6, 49
  See also methane-ice
methane-ice:
  comets **4:** 54
  Neptune **5:** 54
  Pluto **4:** 45
  Triton **5:** 56
Metis (Jupiter moon) **5:** 34, 35
metric (SI)/U.S. standard unit
  conversion table **all**
  volumes: 58
micrometeorites **6:** 55, **7:** 7
microwavelengths **7:** 7
microwave radiation **1:** 54, 56,
  8: 57
microwaves, energy from Sun **2:** 9
military **6:** 16, 17, 18–19, 26; **8:** 6,
  8, 28, 34
Milky Way Galaxy **1:** 4, 7, 11, 18,
  34, 37, 38, 40, 50, 51, **2:** 6–7,
  46
Mimas (Saturn moon) **5:** 36, 40, 42,
  44, 45, 46
minor planets **4:** 10, 47, 48
Mir (space station) **7:** 2, 32, 36–41,
  42, 44, 46, 54
Miranda (Uranus moon) **5:** 51
missiles. *See* ballistic missiles
M number **1:** 12, 13
molecules **1:** 14, 17, 56, **2:** 8, 21, 56,
  3: 22, 24, 44, **4:** 54, **5:** 24
Moon **2:** 50, **3:** 5, 6, 8, 9, 10, 18,
  19, 24–25, 42–57, **4:** 8, 9
  age **3:** 54
  atmosphere **3:** 5, 6, 44, 45, 50,
    52, **4:** 9, **6:** 49
  axis **3:** 8, 9
  centrifugal forces **3:** 10
  density **3:** 54
  formation **2:** 54, **3:** 54, 56–57
  gravity **3:** 8, 10, 42, 44, 51, 54
  heat **3:** 44, 54, 55, 56
  inside **3:** 54–55
    asthenosphere **3:** 55
    core **3:** 54, 55
    crust **3:** 54, 55, 56
    lithosphere **3:** 55
    mantle **3:** 54, 55
    quakes **3:** 55
  landing **6:** 49–51. *See also*
    Apollo (Moon mission);
    Moon rover
  light **3:** 13, 47
  magnetism/magnetic field **3:** 55
  Mars, Moon seen from **3:** 6
  meteorite collisions **3:** 45, 52,
    48–49, 50, 57
  orbit, eccentric **3:** 8, 9, 10, 11,
    12–13, 18, 42, 54
  phases **3:** 9, 12–13
  radiation **3:** 55
  radioactive decay **3:** 44, 54
  reflect, ability to **3:** 47

Moon (*continued...*)
  size **1:** 6, **2:** 14, **3:** 19, 42
  sky, color of **3:** 44
  Sun, fragments in rock **3:** 52
  surface features 45–54
    basins **3:** 42, 46, 56
    craters **2:** 51, **3:** 42, 45, 46, 47,
      50–51, 52, 55, 56, 57, **4:** 9
    erosion **3:** 49, 50
    highlands **3:** 46, 47, 52, 54
    lava/lava flows **3:** 42, 46, 47,
      48–49, 52, 54, 56, **4:** 9
    mare (pl. maria) **3:** 42–43, 46,
      47, 52, 54, 55, **4:** 31
    mascons **3:** 54, 55
    Moon rock **3:** 52–54, 55
    mountains **3:** 50
    "soils" **3:** 44, 45, 49, 52–53
    volcanoes/volcanic activity
      3: 46, 52, 54, 55, 56
  synchronous rotation **3:** 8
  temperature **3:** 44
  tides **3:** 8–9, 10, 18
  tilt **3:** 8, 11
  See also eclipses; Ranger; *and*
    under argon; dust; glass;
    helium; hydrogen; iron;
    light; neon; potassium;
    rock; titanium; zinc
Moon rover **3:** 48–49, **6:** 51
moons (satellites):
  definition **2:** 4
  formation **2:** 54, 57, **3:** 54,
    56–57, **5:** 7, 18
  Galilean moons **5:** 8, 18, 19–34
  gas giant **2:** 50, 57, **5:** 6–7
  irregular satellites **5:** 7, 56
  Jupiter **5:** 6–7, 8, 9, 15, 16, 18–35
  Mars **4:** 42
  Neptune **5:** 6–7, 55–56
  regular satellites **5:** 7
  rocky planet **2:** 57, **4:** 8
  Saturn **5:** 6–7, 36, 40–43, 46
  shepherd satellites **5:** 41, 46, 51
  solar system **2:** 46, 47, 50
  Uranus **5:** 6–7, 48, 50–51
  See also Moon
Moulton, Forest Ray **2:** 53
Mount Everest **4:** 40, **8:** 31
mountains/mountain ranges:
  Earth **3:** 32, 33 34, 35, 36, 38, 41,
    7: 57, **8:** 30, 31, 33, 36, 38,
    42, 44, 46, 54
  Io **5:** 21
  Mercury **4:** 16, 17, 18
  Moon **3:** 50
  Venus **4:** 25, 26, 28, 29
multispectral images **8:** 33

N

NASA (National Aeronautics and
  Space Administration) **2:** 20,
  6: 17, 35, **7:** 23, 44, 46, 53, **8:** 7,
  10, 15, 20, 28, 34, 41, 44
  description **all volumes:** 2
  See also United States space
    initiatives
Navstar (satellite) **8:** 14, 15
nebula (pl. nebulae):
  cosmos **1:** 4
  formation **1:** 20
  Hubble, Edwin Powell **1:** 42
  images of **1:** 10, 11, 12, 13,
    14–15, 16, 18, 25, 26–27, 51,
    5: 4, **4:** 7, **8:** 56
  solar system origin **2:** 53, 55, 56
  See also Butterfly Nebula;
    Cat's Eye Nebula; Crab
    Nebula; Dumbbell Nebula;
    Eagle Nebula; Eskimo
    Nebula; Eta Carinae; gas
    and dust clouds; Lagoon
    Nebula; Orion Nebula;
    Stingray Nebula; *and under*
    ultraviolet radiation

neon:
  Moon **3:** 44
  solar wind **3:** 44
  See also noble gases
Neptune (planet) **2:** 46, 47, 49, 50,
  55, **4:** 5, 57, **5:** 2, 4, 6, 52–57,
  6: 55
  atmosphere **5:** 53, 54
    atmospheric pressure **5:** 54
    clouds **5:** 53, 54, 55
    storms **5:** 53, 54
    weather patterns **5:** 53, 54
    winds **5:** 53, 54
  axis **5:** 53
  centrifugal forces **5:** 53
  composition **5:** 4, 6, 55
  density **5:** 53, 55
  direction **5:** 4
  geysers **5:** 55
  gravity **5:** 5, 56
  Great Dark Spot **5:** 54
  heat **5:** 53, 54
  magnetic field **5:** 55, **6:** 55
  moons **5:** 6–7, 55–56
  orbit **2:** 47, **4:** 44, **5:** 4, 53, **6:** 56
  pressure **5:** 54
  probes. *See* Pioneer; Voyager
  radiation **5:** 53
  rings **5:** 6, 7, 55
  rotation **5:** 53
  Scooter, the **5:** 54
  seasons **5:** 53
  shape and size **5:** 4, 53
  Small Dark Spot **5:** 54
  temperature **5:** 54
  tilt **5:** 53
  See also under ammonia-ice;
    helium; hydrogen; ice,
    water; light; methane;
    methane-ice; reflect,
    ability to; rock
Nereid (Neptune moon) **5:** 55, 56
neutrinos **1:** 31, 54, **2:** 9
neutrons **1:** 30, 31, 34
neutron stars **1:** 21, 31, 33, 34, 35,
  36, **8:** 56
New General Catalogue **1:** 12, 13
New Horizons (probe) **4:** 57
new Moon **3:** 13. *See also* phases
Newton, Sir Isaac **1:** 6, **2:** 38, 53,
  4: 55, **6:** 12, 41
nickel:
  Europa **5:** 29
  meteorites **4:** 6, 57
  stars **1:** 19
Nimbus (satellite) **8:** 12–13, 20
9/11 **8:** 16, 17
nitrogen:
  auroras **2:** 45
  comets **4:** 54
  Earth **3:** 22, **5:** 41
  Mars **4:** 34
  planetary nebula **2:** 8
  Pluto **4:** 44
  spacecraft air supply **6:** 31, **7:** 34
  Sputnik sensors **6:** 23
  stars **1:** 28, 30
  Sun **2:** 17
  Titan **5:** 41, 42
  Triton **5:** 56
  Venus **4:** 23
  See also nitrogen-ice
nitrogen-ice, Triton **5:** 56
nitrogen tetroxide **6:** 11, **7:** 20
noble gases, Mars **4:** 34. *See
  also* neon
Northern Lights **3:** 21. *See
  also* auroras
nova (pl. novae) (stars) **1:** 30.
  See also supernova
nova (pl. novae) (Venus) **4:** 27
nuclear reactions, stars **1:** 23, 24,
  30, **2:** 8, 9, 10, 54. *See also*
  fusion

O

OAO (Orbiting Astronomical
  Observatory) **8:** 56
Oberon (Uranus moon) **5:** 50, 51
Oberth, Hermann **6:** 14, 16
oceans:
  Earth **2:** 10, **3:** 10, 18, 19, 27,
    28, 29, 30, 40, 41, **4:** 26, 36,
    7: 57, **8:** 20, 21, 22
  Europa **5:** 19, 29
  Ganymede **5:** 32
Oceanus Procellarum (Moon)
  3: 42–43, 46
Olympus Mons (Mars volcano)
  4: 40, 41
Oort, Jan Hendrik **4:** 55
Oort cloud **2:** 51, 52, **4:** 55
Ophelia (Uranus moon) **5:** 51
orbiter. *See* Space Shuttle
Orbiting Astronomical Observatory
  (OAO) **8:** 56
Orbiting Solar Observatory (OSO)
  8: 56
orbits:
  explained **2:** 53, 54
  first Earth orbit. *See* Sputnik
  first manned Earth orbits, *see*
    Gagarin, Yuri; Glenn, John
  geostationary **7:** 5, 25, **8:** 13
  geosynchronous **7:** 10, 12–13,
    8: 13, 14–15, 19, 23
  Kuiper belt object **2:** 52
  orbiting velocity **6:** 8, 9, 20,
    7: 10, 11, 12, 13, **8:** 13
  polar **8:** 26, 34
  quaoar **4:** 47
  satellite **1:** 9, **6:** 20, 22–24, 26,
    30, **7:** 5, 6, 12–13; **8:** 4, 6,
    10, 13, 14–15, 30, 34
  spacecraft **4:** 14, **6:** 8, 9, 33,
    34–35, 38, 41, 43, 44, 45,
    46, 47, 49, **7:** 6, 10–11, 13,
    25, 30
  See also deorbit; eccentric
    orbits; trajectory; *and
    under* asteroids; comets;
    Earth; International Space
    Station (ISS); Jupiter; Mars;
    Mercury; Moon; Neptune;
    Pluto; Saturn; stars; Sun;
    Uranus; Venus
Orientale Basin (Moon) **3:** 46
Orion Nebula **1:** 13, 17
OSO (Orbiting Solar Observatory)
  8: 56
outer planets. *See* Jupiter;
  Neptune; Pluto; Saturn;
  Uranus
oxygen:
  auroras **2:** 45
  comets **4:** 54
  Earth **3:** 5, 22, 27, 30, 39, **5:** 41
  Mars **4:** 34
  Mercury **4:** 16
  rocky planets **2:** 50, **4:** 8
  space **2:** 56
  spacecraft air supply **6:** 31, 46,
    7: 34, 41
  spacecraft fuel (liquid oxygen)
    6: 10, 11, 16, 19, 56, **7:** 14,
    15, 18, 20, 21
  stars **1:** 16, 26, 28, 29, **2:** 8
  Sun **2:** 17
ozone **2:** 41, **3:** 22, 24, **8:** 26
ozone hole **8:** 26

P

Pallas (asteroid) **4:** 10, 48, 51
Pan (Saturn moon) **5:** 40, 41, 45
Pandora (Saturn moon) **5:** 40, 45
parachutes **5:** 12, **6:** 26, 27, 33, 40,
  44, 53, **7:** 20, 28, 29
parsec, definition **1:** 11
paths. *See* orbits; trajectory
Pavonis Mons (Mars volcano) **4:** 41

payload **2:** 20, **6:** 7, 8, 9, 10, 11, 16, 18, 20, 22, **7:** 11, 15, 18, 23, 24, 25, 26. *See also* satellites; spacecraft
Payne, Cecilia **2:** 17
penumbra (eclipse) **3:** 15
penumbra (sunspot) **2:** 18, 26–27, 28
perigee **3:** 8, 14
phases:
   Earth **3:** 6
   Mercury **4:** 14
   Moon **3:** 9, 12–13
   Venus **4:** 22
Phobos (Mars moon) **4:** 42
Phoebe (Saturn Moon) **5:** 41, 45
phosphorus:
   Jupiter **5:** 15
   Saturn **5:** 40
photoevaporation **1:** 16
photons **1:** 17, 36, 46, **2:** 18, 21, 23, 38
photosphere **2:** 19, 22, 23–26, 28, 32, 33, 35, 37
photosynthesis **3:** 27
Pioneer (probe) **5:** 8, 44, **6:** 52, 54, 55, 57
Pirs (universal docking module, ISS) **7:** 44, 52, 53
planet, origin of word **1:** 9
planetary nebula **1:** 13, 21, 26–28, 29, 30 31, 33, **2:** 8, 11
planetesimals **4:** 11
planet formation:
   Earth **3:** 30, 39
   explained **2:** 49, 52–54, 56–57, **4:** 7, 8, 11, **5:** 4, 6, 34
   Jupiter **2:** 56–57, **5:** 4–5, 17, 18
   Mars **4:** 43
   other solar systems **1:** 48, 49, 57
   Saturn **5:** 17
   Uranus **2:** 54, **5:** 48, 49, 50
planets. *See* Earth; Jupiter; Mars; Mercury; Neptune; planet, origin of word; planet formation; Pluto; Saturn; Uranus; Venus
plant life **3:** 16, 27, 29, 30, **7:** 55
plasma:
   black hole **1:** 39
   comets **4:** 53
   probe **6:** 55
   Sun (solar wind) **2:** 16, 22, 26, 32, 33, 35, 38, 42, 45, 46
   See also coronal mass ejections; solar wind
plates/plate boundaries **3:** 30–37, 39, 41
Pluto **2:** 46, 47, 48, 49, 50, 51, 52, 55, **4:** 5, 7, 8, 44–47, 57, **5:** 4, 56
   atmosphere **4:** 46
   axis **4:** 45
   brightness **4:** 45
   composition **2:** 46, **4:** 44, 45
   density **4:** 45
   direction, retrograde **4:** 45
   heat **4:** 46
   moon **4:** 44, 46–47, 57
   orbit, eccentric **2:** 47, 48, 52, **4:** 7, 44, 46
   pressure **4:** 46
   rotation **4:** 45
   shape and size **4:** 5, 45
   surface features:
     basins **4:** 45
     craters **4:** 45
     polar caps **4:** 45
   synchronous rotation **4:** 46
   temperature **4:** 46
   See also under ice, water; methane; methane-ice; nitrogen; reflect, ability to; rock
polar caps:
   Ganymede **5:** 30–31
   Jupiter **5:** 16

polar caps (*continued...*)
   Mars **4:** 31, 34
   Pluto **4:** 45
   See also ice caps
pollution **7:** 57, **8:** 36, 37
Polyakov, Valery **7:** 39
Population I and II stars **1:** 19, 30
potassium:
   Mercury **4:** 16
   Moon **3:** 44
probes, space **2:** 52, **3:** 6, **6:** 4, 9, 52–57, **8:** 4. *See also* Cassini-Huygens; Galileo; Magellan; Mariner; Mars Global Surveyor; New Horizons; Pioneer; Stardust; Venera; Voyager
Progress (unmanned ferry) **7:** 32, 38, 39
projectiles. *See* rockets
Prometheus (Io volcano) **5:** 22
Prometheus (Saturn moon) **5:** 40, 45
prominences **2:** 15, 19, 33–35
propellants **6:** 7, 8, 10, 11, 12, 14, 16, 19, 20, **7:** 20, 21, 26, 33
propulsion systems **6:** 7, 9, 10, 11, 43, 55, **7:** 32. *See also* engines
Proton (rocket) **7:** 51, 52
protons **1:** 30, 34, **2:** 8, 9, 32, 33, 41, 42, 45, **3:** 19
protostars **1:** 18
Proxima Centauri (star) **1:** 11, **2:** 6, 13
pulsars **1:** 21, 33, 34, 35

Q

quantum theory **7:** 55
Quaoar (minor planet) **4:** 47
quasars **1:** 33, 35, 36, 38, **8:** 56

R

radar **7:** 26, **8:** 17, 23, 45–53
   Venus **4:** 20–21, 25, **6:** 54
   See also Shuttle Radar Topography Mission (SRTM); surveying (and mapping) Earth
radiation:
   Big Bang **1:** 54, 56
   black holes **1:** 38, 39
   comets **4:** 53
   cosmos **1:** 5
   detecting **1:** 7
   Earth **4:** 36, **8:** 26, 57
   Jupiter **5:** 15
   Mars **4:** 36
   Moon **3:** 55
   Neptune **5:** 53
   probes **6:** 55
   quasars **1:** 35
   Saturn **5:** 38
   solar (Sun) **2:** 8, 9, 10, 16–17, 22, 41, **3:** 29, **4:** 24, **5:** 6, 12, **6:** 24, **8:** 8, 26
   space **1:** 7, **2:** 8, **7:** 7
   stars **1:** 14, 20, 21, 23, 29
   universe **1:** 7, 20, 54
   Uranus **5:** 50
   Venus **4:** 25
   See also infrared; microwave radiation; synchrotron radiation; ultraviolet radiation; Van Allen belts
radiation belts. *See* Van Allen belts
radiative zone **2:** 18–19, 21, 22
radio galaxies **1:** 45
radio interference **2:** 38, 45
radioisotopes:
   power source **6:** 54, 55
   rock dating **3:** 54
radio stars **1:** 36, 45
radio telescopes **1:** 4, **2:** 8
radio waves:
   black holes **1:** 36
   Earth (ionosphere) **2:** 41, **3:** 22, **6:** 23

radio waves (*continued...*)
   Jupiter **5:** 8, 15
   pulsars **1:** 21, 34
   quasars **1:** 35
   Sun **2:** 9, 32, 46
   universe **1:** 7
   See also radar
rain (precipitation) **3:** 23, 29, **8:** 21, 22, 23, 25
Ranger (spacecraft) **6:** 53
rays, crater **3:** 46, 50, 51
reaction **2:** 38, **6:** 6, 7, 12, 41. *See also* laws of motion
red giants **1:** 21, 26, 28, 30, **2:** 10
"red planet" **4:** 6, 30
red spot (Saturn) **5:** 40
Redstone (rocket) **6:** 34, 35
reflect, ability to:
   dust and gas **8:** 54
   Earth clouds **3:** 26, 29
   Enceladus **5:** 42
   galaxies **1:** 40
   Mars **4:** 31
   Mercury **4:** 16
   Moon **3:** 47
   Neptune **5:** 53
   Pluto **4:** 45
   Triton **5:** 56
   Uranus **5:** 49
   Venus clouds **4:** 24, 25
   See also radar
relativity, theory of **1:** 46, 48, **2:** 9
remote manipulator system (RMS):
   International Space Station (ISS) **7:** 45, 47, 49, 52
   Space Shuttle **7:** 24, **8:** 10
remote sensing **8:** 30. *See also* surveying (and mapping) Earth
resonance **4:** 49
retroburn **6:** 45
retrograde direction:
   Pluto **4:** 45
   Triton **5:** 56
   Venus **4:** 20
retrorockets/retrofiring **6:** 31, 33, 38, 40, 43, **7:** 20, 27. *See also* retroburn
Rhea (Saturn moon) **5:** 36, 41, 45
rifts/rift valleys:
   Ariel **5:** 50
   Earth **3:** 33, **4:** 26, 41
   Mars **4:** 41
   Venus **4:** 26–27
   See also faults/fractures
rings, planet. *See under* Jupiter; Neptune; Saturn; Uranus
robotic arm. *See* remote manipulator system (RMS)
rock **2:** 56
   Callisto **5:** 30, 34
   comets **2:** 46, **4:** 54
   Earth **2:** 10, **3:** 6, 23, 28, 29, 30, 32, 34, 36, 39, 40, **4:** 41, **8:** 44, 45
   Europa **5:** 29
   Ganymede **5:** 6, 18, 30, 32
   Gaspra **4:** 10
   isotope dating **3:** 54
   Io **5:** 21, 26
   Jupiter **5:** 17
   Mars **4:** 36–37, 43
   Moon **3:** 6, 42, 44, 48–49, 52–54, 55, 56
   Neptune **5:** 6, 55
   Pluto **2:** 50, **4:** 8, 44, 45
   rocky bodies **4:** 4, 5, 6–7, 10–11, **2:** 46. *See also* asteroids; comets; interplanetary dust; meteoroids
   Saturn/Saturn's rings **5:** 7, 38, 45
   solar system **2:** 46, 56, 57
   Tethys **5:** 42
   Triton **5:** 56
   Uranus/Uranus' moons **5:** 6, 48, 49, 50, 51

rock (*continued...*)
   Venus **4:** 22, 25, 27, 29
   See also gas giant moons; lava/lava flows; rock cycle; rocky planet moons; rocky planets; sedimentary rock
rock cycle **3:** 29, 40–41
rocketry **6:** 14–21. *See also* rockets
rockets **4:** 52, **6:** 6–21, 24, 26, 27, 31, 34, 35, 40, 41, 42–43, 44, 45, 46, 49, 56, **7:** 14, 17, 22, **8:** 6–7, 10, 13, 14, 29–30
   aerodynamic design **6:** 7, 16, 22, **7:** 17
   lift/lifting force **6:** 9, 24, 46, **7:** 9, 14, 17, **8:** 10
   See also Agena (rocket); booster pods; engines; launchers/launch vehicles; propellants; propulsion systems; Proton (rocket); Redstone (rocket); retrorockets; rocketry; takeoffs; thrusters; Viking program (rockets)
rocky planet moons **2:** 57. *See also* Moon; *and under* Earth; Mars; Mercury; Pluto
rocky planets **2:** 50, **4:** 4 *and throughout*, **5:** 6
   formation **2:** 56–57, **4:** 7, 8
   See also Earth; Mars; Mercury; Pluto; Venus
Russell, Henry Norris **2:** 10
Russia **6:** 14, 18, **7:** 4, 30, 44, 45, 52, **8:** 25, 37, 38
Russian Space Agency **7:** 52
Russian space initiatives **3:** 54, **6:** 18, 30, **7:** 4, 30, 44, 46, **8:** 9. *See also* International Space Station (ISS); Mir; Salyut; Soyuz; Sputnik; Venera; Vostok

S

Salyut (space station) **7:** 30–32, 33, 36, 38, 42
sand dunes:
   Earth **8:** 36
   Mars **4:** 36
   Venus **4:** 24
Saros cycle **3:** 14
satellites (man-made) **1:** 9, **2:** 20–21, 42, **6:** 8, 9, 17, 20, 22–29, 30, **7:** 4, 5, 6–7, 16, 17, 25, **8:** 4 *and throughout*
   astronomical **8:** 54–57
   communications **6:** 28–29, **7:** 5, 6–7, 12, 13, **8:** 4, 7, 8, 13, 14–15, 17
   cost **8:** 6–7, 10, 15
   design **6:** 23, 28, **7:** 6–7, 9
   formation **8:** 14–15
   geostationary **6:** 28, **8:** 7, 14, 23, 26
   geosynchronous **7:** 12, 13
   global positioning system (GPS) **6:** 23, **7:** 5, **8:** 4, 7, 13, 14
   inertial upper stage (Space Shuttle) **7:** 25
   military **8:** 6, 8, 34
   polar-orbiting **7:** 13, **8:** 13, 26, 34
   reflecting **8:** 8, 9
   telephone **8:** 4, 7, 8
   television **8:** 4, 7
   weather **6:** 27–28, **7:** 12, **8:** 4, 14, 17, 20–27
   See also Aqua; COBE; Discoverer; Echo; ERTS-1; Explorer; GOES; Grace; Hubble Space Telescope; INTELSAT VI; Landsat; Meteosat; Navstar; Nimbus; SOHO; Sputnik; Syncom;

satellites (man-made) (continued...)
    Terra; Telstar; Tiros;
    and under centrifugal
    forces; infrared
satellites (natural). See moons
    (satellites)
Saturn (launcher) 6: 4–5, 11, 17, 20,
    43, 7: 33, 34
Saturn (planet) 1: 9, 2: 46, 47, 49,
    50, 55, 3: 19, 4: 4, 5: 2, 4, 5,
    6–7, 17, 36–47, 6: 52, 54, 55,
    56, 57
    atmosphere 5: 37, 38, 39–40
        clouds 5: 38, 39, 40
        weather patterns 5: 38
        winds 5: 38, 39
    axis 5: 37
    centrifugal forces 5: 39
    composition 5: 4, 6, 37–39
    density 2: 50, 5: 37
    direction 5: 4
    formation 5: 17
    gravity/gravitational field 5: 6,
        39, 45, 46
    heat 5: 38
    magnetic field 5: 37, 38, 6: 55
    mass 5: 37
    moons 5: 6–7, 36, 40–43, 46
    orbit 2: 47; 5: 4, 36, 6: 56
    pressure 5: 38, 39
    probes. See Cassini-Huygens;
        Pioneer; Voyager
    radiation 5: 38
    radius/radii (Rs) 5: 45
    red spot 5: 40
    rings 5: 6, 7, 36–37, 40, 44–47
    rotation 5: 37
    shape and size 3: 19, 5: 5, 36, 37,
        38, 39
    star, potential to be 5: 37
    temperature 5: 39
    tilt 5: 37
    See also under ammonia;
        ammonia-ice; helium;
        hydrogen; ice, water;
        metallic hydrogen;
        methane; phosphorus; rock
Saturn radii (Rs) 5: 45
Schirra, Jr., Walter M. 6: 35
Scooter, the (Neptune) 5: 54
seasons:
    Earth 3: 10–11, 8: 19
    Mars 4: 32
    Neptune 5: 53
sediment 3: 34, 41, 8: 41
sedimentary rock 3: 29, 41
seismic waves:
    earthquake 3: 38
    "moon" quakes 3: 55
sensors 1: 7, 6: 23, 8: 8, 15, 16–19,
    20, 24, 25, 26, 33, 34, 37, 54, 56
Shepard, Alan B. 6: 34–35
shepherd satellites 5: 41, 46, 51
shock waves:
    Big Bang 1: 56
    bow 2: 40, 41, 42
    solar 2: 32, 33, 41
Shoemaker-Levy 9 (comet) 4: 53
shooting stars. See meteoroids;
    meteors
Shuttle. See Space Shuttle
Shuttle Radar Topography Mission
    (SRTM) 8: 46–53
sidereal month 3: 8–9
Sif Mons (Venus) 4: 28, 29
silicate minerals:
    Earth 3: 39
    meteorites 4: 57
    "stardust" 4: 10
silica tiles (ISS) 7: 28
silicon:
    Earth 3: 39
    rocky planets 4: 8
    space 2: 56
    stars 1: 14, 30
    Sun 2: 17

sky, picture of entire 1: 4–5
Skylab (space station) 2: 31,
    7: 30-31, 33–35, 36, 42, 44
slingshot/slingshot trajectory 4: 14,
    54, 6: 52, 55, 56, 57
Small Dark Spot (Neptune) 5: 54
snow:
    Earth 3: 6, 23, 28, 29, 8: 25,
        30, 38
    Mars 4: 34
sodium:
    Mercury 4: 16
    Sun 2: 16
    universe 1: 14
SOHO (Solar and Heliospheric
    Observatory) satellite 2: 20–21
"soils":
    Earth 3: 39, 40
    Gaspra (asteroid) 4: 10
    Mars 4: 34, 43
    Mercury 4: 17
    Moon 3: 44, 45, 49, 52–53
    Venus 4: 25
solar arrays. See solar panels
solar cells 6: 25, 26, 28, 29, 7: 50,
    8: 19. See also solar panels
solar eclipses 3: 14, 15
solar flares 2: 19, 24, 32–33, 35, 45
solar nebula 2: 54
solar panels (arrays) 2: 20, 6: 52,
    7: 6, 7, 32, 33, 34, 35, 36, 38,
    42, 52, 8: 14, 20, 57. See also
    solar cells
solar prominences 2: 15, 19, 33–35
solar quake 2: 32
solar radiation. See under radiation
solar sail 2: 38–39
solar system 2: 46–57, 4: 4, 5: 4
    age 3: 54, 4: 11, 5: 4
    composition 2: 4, 46, 49–52,
        3: 44, 4: 54
    diagram 2: 48–49, 4: 4–5
    formation 2: 10, 52–57, 4: 10, 11
    mass 2: 4, 14, 49
    organization 2: 48, 49
solar wind 2: 4, 9, 11, 20, 33, 37,
    38–45, 46, 3: 20, 21, 24, 44, 52,
    6: 54, 7: 7, 56, 8: 56
    comets 4: 54
    discovery 4: 54
    heliosphere 2: 40, 41
    Mercury 4: 16
    Moon 3: 44, 52
    Uranus 5: 50
    Venus 4: 22
    See also auroras;
        magnetospheres; stellar
        wind; and under neon
sonic boom 4: 57, 7: 29
Southern Cross (Crux) 1: 8
Southern Lights 3: 21. See also
    auroras
South Pole-Aitken Basin (Moon)
    3: 46
Soviet space initiatives. See Russian
    space initiatives
Soviet Union, former 6: 14, 18, 19,
    22, 24, 26, 30, 33, 38, 7: 4, 30,
    32, 44, 8: 8
Soyuz (ferry) 6: 18, 33, 7: 32–33, 38,
    39, 44, 45
space, Sun's corona 2: 35
spacecraft 2: 20, 38, 6: 4 and
    throughout, 7: 4 and
    throughout
    capsules 6: 26, 27, 30, 33, 34, 35,
        37, 38, 41, 44, 51, 7: 32, 35
    design 7: 6–7, 9, 17
    ejection seats 6: 31, 33, 40
    future development 7: 56
    gimbals 6: 12, 13, 16, 7: 20, 25
    hatches, entry and exit 6: 40, 41,
        7: 56–57
    heat shields 5: 12, 6: 26, 33, 40,
        44, 46, 51, 7: 27, 8: 57
    nose cones 6: 7, 10, 22

spacecraft (continued...)
    recovering 6: 26, 30, 35, 41, 51,
        7: 20, 23, 8: 15
    trusses 6: 52, 7: 45, 52, 53
    See also Apollo (Moon
        mission); docking,
        spacecraft; engines; fuel,
        spacecraft; Gemini; lunar
        module; Mercury; probes,
        space; Progess; Ranger;
        retrorockets; rockets;
        Space Shuttle; Viking
        mission (Mars); Vostok;
        and under centrifugal
        forces; hydrogen; nitrogen;
        oxygen
spaceflight, manned. See manned
    spaceflight
Spacehab 7: 24
Spacelab 7: 24
space probes. See probes, space
space race 6: 19, 22–29, 7: 4
Space Shuttle (Space
    Transportation System (STS))
    6: 11, 13, 7: 4, 8, 9, 13, 14–29,
    36–37, 38, 40–41, 44, 46, 8: 4,
    10, 45, 48–49, 50
    boosters 6: 11, 7: 8–9, 14, 15, 17,
        18–20, 21
    external tank 6: 11, 7: 8–9, 14,
        15, 17, 18, 19, 20, 21
    orbiter 6: 11, 7: 14, 15, 17, 18,
        19, 20–23, 24, 25, 26, 27–29
    See also Atlantis; Challenger;
        Columbia; Discovery;
        Endeavour; Enterprise;
        inertial upper stage;
        Shuttle Radar Topography
        Mission (SRTM)
space stations 7: 4, 6, 13, 17,
    30, 32, 36, 42. See also
    International Space Station
    (ISS); Mir; Salyut; Skylab
Space Transportation System (STS).
    See Space Shuttle
spacewalking. See extravehicular
    activity
spicules 2: 33, 35
Sputnik (satellite) 6: 18, 22–24, 26,
    8: 8, 9, 30
SRTM (Shuttle Radar Topography
    Mission) 8: 46–47, 48, 50,
    52–53
Stardust (probe) 4: 54
"stardust." See interplanetary dust
stars 1: 7, 8–9, 14–39
    atmosphere 1: 26, 2: 53
    brightness 1: 18, 21, 24, 30, 33.
        See also magnitude
    cataloguing 1: 12
    classification 1: 19
    clusters (open and globular)
        1: 12, 19–20, 8: 57
    elements. See under calcium;
        carbon; helium; hydrogen;
        iron; nickel; nitrogen;
        oxygen; silicon; sulfur
    energy 1: 21, 23, 26, 30, 31,
        35, 45
    formation 1: 14–18, 38
    fuel 1: 21, 23, 24, 26, 29, 2: 8–9
    gravity/gravitational field 1: 18,
        19, 20, 23, 24, 26, 29, 30,
        31, 42, 56, 2: 8, 9
    heat 1: 18, 20, 21, 23, 26, 29
    life cycle 1: 14–15, 19, 20–33,
        2: 10–11, 8: 57
    light 1: 16, 18, 24, 29, 30, 33,
        39, 42, 45
    magnetic field 1: 34
    magnitude 2: 13, 8: 56
    near-star event 2: 53
    nuclear reactions 1: 23, 24, 30,
        2: 8, 9, 10, 54
    orbit 1: 42
    pressure 1: 18

stars (continued...)
    radiation 1: 14, 20, 21, 23, 29
    rotation in galaxy 1: 42
    temperature 1: 21
    twinkling effect 8: 54
    See also Betelgeuse (star); binary
        stars; black dwarf stars;
        blue giants; DG Tau (star);
        dwarf stars; main-sequence
        stars; neutron stars; nova;
        Population I and II stars;
        protostars; Proxima Centauri
        (star); pulsars; radio stars;
        Sun; supernova; T-Tauri
        stars; white dwarf stars; and
        under infrared; ultraviolet
        light
stellar winds 1: 14, 18, 25, 26, 28,
    51, 2: 40. See also solar wind
Stingray Nebula 2: 8
stratosphere 3: 22, 24
STS. See Space Shuttle
subduction zones 3: 32, 33, 34,
    36–37
sulfur:
    comets 4: 54
    Earth 3: 39
    Io 5: 21, 23, 24
    Jupiter 5: 14, 15
    stars 1: 16, 30
sulfur dioxide, Io 5: 22, 26
sulfuric acid clouds, Venus 4: 20,
    23, 24
Sun 1: 5, 6, 7, 11, 19, 24, 42, 2: 2, 4,
    6, 7, 8–9, 10–11, 12–45, 46, 48,
    53, 54, 4: 57, 6: 54
    age 2: 10
    atmosphere 2: 12, 18, 20, 22–37
        chromosphere 2: 18, 22, 32,
            33, 35
        corona 2: 19, 22, 35–37, 38,
            41, 53, 3: 15
        coronal loops 2: 35, 36–37
        coronal mass ejections 2: 4–5,
            22–23, 35, 44
        flares 2: 19, 24, 32–33, 35, 45
        photosphere 2: 19, 22, 23–26,
            28, 32, 33, 35, 57
        prominences 2: 15, 19, 33–35
        spicules 2: 33, 35
        storms 2: 33, 42–43, 44–45
    axis 2: 13, 24, 47
    birth 2: 8, 10
    brightness 1: 24, 2: 9, 10, 11,
        12–13, 16
    death 2: 10–11
    density 2: 11, 16, 22, 23
    "diamond ring" 3: 14–15
    Earth, distance from 2: 12
    elements 2: 16–17. See also
        under calcium; carbon;
        helium; hydrogen; iron;
        magnesium; nitrogen;
        oxygen; silicon; sodium
    energy 2: 8–9, 16, 21, 22, 28, 31,
        32, 33, 38, 46, 3: 22, 28, 8: 26
    fuel 2: 8–9
    gravity 2: 8, 9, 14, 16, 46, 3: 10,
        8: 56
    heat 2: 21, 22, 32, 41, 46, 54
    heliopause 2: 40, 41
    helioseismology 2: 16, 32
    heliosphere 2: 40–41, 44, 45
    heliotail 2: 40
    inside:
        convective zone 2: 18–19,
            22, 23
        core 8: 16, 18–19, 21
        radiative zone 2: 18–19, 21, 22
    light 2: 6, 9, 11, 16–17, 21, 23,
        24, 32, 41, 46, 3: 11, 13,
        14–15, 22
    magnetism/magnetic field 2: 16,
        28, 32, 33, 35, 37, 41, 42–45,
        46
    mass 2: 4, 9, 14, 22

71

Sun (continued...)
  modeling to test theories **2:** 16
  orbit **1:** 37, **8:** 26
  pressure **2:** 8, 16
  probes **6:** 54. See also Orbiting Solar Observatory; SOHO satellite
  quakes and vibrations **2:** 16, 32
  radiation, solar **2:** 8, 9, 10, 16–17, 22, 41, **3:** 29, **4:** 24, **5:** 6, 12, **6:** 24, **8:** 8, 26
  radio waves **2:** 9, 32, 46
  radius **2:** 16, 21
  rotation **2:** 12–13, 24, 26
  size **1:** 6, **2:** 14, 15, 36–37
  star type **1:** 19, **2:** 12
  surface features:
    granulation **2:** 24, 26–27, 28
    sunspots **2:** 12–13, 26–32, 35
    temperature **2:** 10, 12, 22, 23, 26, 28, 32, 35, 40
  tilt **2:** 13
  See also auroras; eclipses; solar sail; solar wind; tides; and under convection currents; infrared; plasma; ultraviolet light; ultraviolet radiation
sunspots **2:** 12–13, 26–32, 35
supernova (pl. supernovae) **1:** 13, 21, 30, 31–33, 34, 51, **8:** 54–55, 56
solar system origin **2:** 54
surveying (and mapping) Earth **6:** 23, **7:** 4, 13, 26, **8:** 13, 14. See also global positioning system; Landsat; radar; remote sensing; Shuttle Radar Topography Mission (SRTM); triangulation
synchronous orbit, Charon **4:** 46
synchronous rotation:
  Charon **4:** 46
  Io **5:** 26
  Moon **3:** 8
  Venus **4:** 22
synchrotron radiation **1:** 39
Syncom (satellite) **8:** 14
synodic month **3:** 9
synodic period **4:** 15

T
takeoffs **6:** 4, 11, 24, **7:** 9, 10–11, 13
  Apollo **6:** 10, 42–43, 44, 45, 47
  Cassini/Titan **6:** 21
  Friendship 7 (Mercury) **6:** 36
  Gemini **6:** 38
  Proton (International Space Station) **7:** 51
  Space Shuttle **7:** 14, 17, 18–19, 20, 21, 22, 23, 26–27, 29, **8:** 9
  See also rockets
tectonic plates. See plates
telecommunications **7:** 56. See also telephone
telephone **8:** 4, 7, 8. See also telecommunications
telescopes **1:** 6, 7, 12, **4:** 49, **5:** 18, **7:** 34, **8:** 15, 54–57. See also Hubble Space Telescope; radio telescopes
Telesto (Saturn moon) **5:** 40, 41
television (satellite) **6:** 25, 27, 28, 29, **8:** 4, 7, 13, 14. See also Telstar; Tiros
Telstar (satellite) **6:** 28–29, **8:** 8
Tereshkova, Valentina **6:** 35
Terra (satellite) **8:** 14–15, 27, 41, 44
Tethys (Saturn moon) **5:** 36, 40, 42, 43, 45
Tharsis (Mars) **4:** 41
Thebe (Jupiter moon) **5:** 34, 35
theory of relativity **1:** 46, 48, **2:** 9
thermal imaging **8:** 40–44
thermosphere **3:** 22, 24
Thor (launcher) **6:** 19

thrusters (orientation rockets) **6:** 40, **7:** 38, 39
tidal effect **5:** 22
tides:
  Earth **3:** 8–9, 10, 18
  Moon **3:** 18
time machine, light as **1:** 50, 53
Tiros (satellite) **6:** 27–28, **8:** 17, 20, 26
Titan (launcher) **6:** 19, 21, 38, 56
Titan (Saturn moon) **2:** 50, **5:** 36, 41–42, 45, **6:** 57
Titania (Uranus moon) **5:** 50, 51
titanium:
  Mercury **4:** 18
  Moon **3:** 52
Titov, German **6:** 33
topography **5:** 41, **8:** 43–50. See also Shuttle Radar Topography Mission (SRTM)
total eclipse **2:** 33, **3:** 14
trajectory **6:** 9, 34, 43, 44, 47, 52, 56, **7:** 10–11, 20. See also slingshot trajectory
transponders **8:** 14, 17, 19
triangulation **8:** 48–49, 51
Triton (Neptune moon) **4:** 47, **5:** 55, 56, 57
tropical cyclones **8:** 22. See also hurricanes
troposphere **3:** 22, 24, 28
trusses **6:** 52, **7:** 45, 52, 53
Tsiolkovsky, Konstantin Eduardovich **6:** 14, 16
T-Tauri stars **1:** 18, 48
Tycho Brahe **1:** 6
Tycho impact basin **3:** 42–43
typhoons. See hurricanes

U
ultraviolet light:
  satellite sensors **8:** 26, 56
  stars **1:** 16, 17, 22–23
  Sun **2:** 9, 16–17, 32, 37, 41
  See also photons
ultraviolet radiation:
  Earth **3:** 22, 24
  nebula **1:** 51
  Sun **2:** 33, 41
umbra (eclipse) **3:** 15
umbra (sunspot) **2:** 18, 26–27, 28
Umbriel (Uranus moon) **5:** 50, 51
United States, first photo map **8:** 28–29
United States space initiatives **6:** 24–26, 30, **7:** 4, 44, 45, 46. See also Apollo; Gemini; GOES; International Space Station; Mercury; NASA; probes, space; Skylab
Unity (ISS) **7:** 42–43, 45, 50, 51, 52
universe **1:** 4, 6, 40, 46–57, **2:** 6
  age **1:** 20, 42, 54
  composition **1:** 14, 17, 19, **2:** 6, 17, 56, **4:** 54. See also under calcium; helium; hydrogen; sodium; water
  continuation **1:** 57
  expanding **1:** 42, 53–57
  "observable" **1:** 53
  origin **1:** 6, 42, 54, 55, 56, **8:** 57
  structure **1:** 49–50
  See also Cosmological Principle; and under energy; light; radiation; radio waves; X-rays
Uranus (planet) **2:** 46, 47, 48, 49, 50, 55, **5:** 2, 4, 6, 48–51, **6:** 55
  atmosphere **5:** 49
    auroras **5:** 50
    clouds **5:** 48
    winds **5:** 49
  axis **5:** 48, 49
  centrifugal forces **5:** 49
  color **5:** 49
  density **5:** 48

Uranus (continued...)
  direction **5:** 4, 48
  formation **5:** 48, 49, 50, **2:** 54
  gravitational field **5:** 49, 50
  heat **5:** 50
  inside **5:** 4, 6, 49, 50
  magnetic field **5:** 49, 50, **6:** 55
  mass **5:** 50
  moons **5:** 6–7, 48, 50–51
  orbit **2:** 47, **4:** 44, **5:** 4, 48, **6:** 56
  pressure **5:** 50
  probes. See Pioneer, Voyager
  radiation **5:** 50
  rings **5:** 6, 7, 48, 50, 51
  rotation **5:** 48, 49
  shape and size **5:** 4, 48, 49
  tilt, extreme **2:** 48, 54
  See also under ammonia; ammonia-ice; energy; helium; hydrogen; ice, water; methane; reflect, ability to; rock; water
Ursa Major (Great Bear) **1:** 8
U.S. standard unit/metric (SI) conversion table **all volumes: 58**

V
V1/V2 (Vengeance weapons) **6:** 16, **8:** 29
vacuum **2:** 23, **7:** 7, 17, 56
Valles Marineris (Mars) **4:** 39, 41
Van Allen, Dr. James **6:** 24
Van Allen belts **3:** 19, 24, **6:** 24
Vanguard (launcher) **6:** 19, 24
Vanguard (satellite) **6:** 24, 26
Venera (probe) **6:** 53
Venus (planet) **2:** 10, 42, 46, 47, 48, 50, 55, **4:** 2, 4, 5, 6, 20–29, **6:** 52, 53, 54, 56
  atmosphere **3:** 27, **4:** 20, 23–25, 29
    atmospheric pressure **4:** 23
    clouds, sulfuric acid **4:** 20, 22, 23, 24, 25
    "greenhouse" effect **4:** 24–25
    winds **4:** 24
  axis **4:** 20
  brightness **4:** 20, 24
  centrifugal forces **4:** 22
  death **2:** 10
  density **4:** 22
  direction, retrograde **4:** 20
  heat **4:** 25, **6:** 53
  inside:
    convection currents **4:** 29
    core **4:** 29
    crust **4:** 26, 27, 28, 29
    mantle **4:** 29
    plates **4:** 26, 29
  mass **4:** 22
  orbit **2:** 47, **4:** 7, 20, 22, **6:** 57
  probes. See Magellan; Mariner; Pioneer; Venera
  radiation **4:** 25
  radioactive decay **4:** 29
  rotation **4:** 20, 22
  shape and size **4:** 5, 22
  surface features **4:** 20–21, 25–29
    basins **4:** 26
    calderas **4:** 29
    chasm **4:** 26
    corona **4:** 27
    craters **4:** 25, 28, 29
    depressions **4:** 29
    faults/fractures **4:** 26–27, 28
    highlands **4:** 26
    lava/lava flows **4:** 25, 26, 27, 28, 29
    mountains **4:** 25, 26, 28, 29
    nova **4:** 27
    plains **4:** 25, 26, 27
    plateaus **4:** 26
    ridges **4:** 25, 28
    rifts/rift valleys **4:** 26–27
    sand dunes **4:** 24

Venus (continued...)
    scarp **4:** 26–27
    "soils" **4:** 25
  volcanoes/volcanic activity **4:** 25, 27, 28–29
  synchronous rotation **4:** 22
  temperature **4:** 25
  See also under carbon dioxide; magma; nitrogen; phases; radar; rock
Vesta (asteroid) **4:** 10, 48, 51
Viking mission (Mars) **4:** 36–37, 38–39
Viking program (rockets) **8:** 29
volcanoes/volcanic activity:
  Earth **3:** 27, 30, 32, 33, 38, 39, 40, 41, **7:** 57, **8:** 38–39, 46–47
  Enceladus **5:** 6–7, 42
  Io **2:** 50, **5:** 15, 19, 20–24, 26, 35. See also Prometheus
  Mars **4:** 38, 40, 41, 43. See also Albor Tholus; Ascraeus Mons; Ascraeus Mons; Elysium Mons; Hecates Tholus; Olympus Mons; Pavonis Mons
  Moon **3:** 46, 52, 54, 55, 56
  Venus **4:** 25, 27, 28–29
von Braun, Wernher **6:** 16–17, **7:** 42
Vostok **6:** 18, 30, 31, 33, 34, 35
Voyager (space probe) **2:** 6, **5:** 8, 18, 19, 21, 26–27, 36, 44, 45, 46, 49, 50, 55, **6:** 52, 54–56

W
water:
  Callisto **5:** 18, 19
  Earth **3:** 16, 22, 23, 27, 28, 29
  Europa **5:** 29
  Ganymede **5:** 18, 19, 30, 32
  Jupiter **5:** 15
  Mars **4:** 31, 33, 34, 40, 41
  Miranda **5:** 51
  solar system origin **2:** 56–57
  universe, most common molecule in **2:** 56
  Uranus **5:** 48, 49
  See also ice, water; water cycle; water vapor
water cycle **3:** 24, 28, 29, 32, 36, 40
water-ice. See ice, water
water vapor:
  comets **4:** 53
  Earth **3:** 22, 23, 28, 29, 39, **8:** 2, 22, 24, 25
  Mars **4:** 32, 34, 36, 40
weapons, atomic **6:** 18. See also ballistic missiles
weather satellites **6:** 27–28, **7:** 12, **8:** 4, 14, 17, 20–27
weathering **3:** 46, 50, **4:** 25. See also erosion
weightlessness **6:** 16, **7:** 42, 57
Whirlpool Galaxy **1:** 44–45
white dwarf stars **1:** 21, 29, 32, **2:** 11, **3:** 33
woman, first in space **6:** 35

X
X-rays:
  black holes **1:** 35, 36, **8:** 56
  cosmos **1:** 5, 7
  galaxies **1:** 45
  neutron stars **8:** 56
  pulsars **1:** 34
  quasars **1:** 35
  space, first detection in **8:** 56
  Sun **2:** 9, 32, 33, 41
  supernova **8:** 56
  universe **1:** 5, 7

Z
Zarya (ISS) **7:** 42–43, 45, 50, 51, 52
zinc, Moon **3:** 52
Zvezda service module (ISS) **7:** 44, 52